ENVIRONMENTAL GEOGRAPHY AND NATURAL HAZARDS

CONCEPT'S INTERNATIONAL SERIES
IN
ENVIRONMENT
(CISE)

General Editor : K.V. PALIWAL

CISE-1 Environmental Geography and Natural Hazards : Exigencies of Appraisal in Highland – Lowland Interactive Systems by **A.A. Pirazizy**
Rs. 250

CISE-2 Industrial Waste Disposal and Sewage Irrigation : An Annotated Bibliography by **K.V. Paliwal**
(in press)

Concept's International Series in Environment-1

General Editor: **K.V. Paliwal**

ENVIRONMENTAL GEOGRAPHY AND NATURAL HAZARDS

Exigencies of Appraisal in Highland – Lowland Interactive Systems

A. A. PIRAZIZY

CONCEPT PUBLISHING COMPANY, NEW DELHI 110059

ISBN 81-7022-424-1

First Published 1992

Published and Printed by
Ashok Kumar Mittal
Concept Publishing Company
A/15-16, Commercial Block, Mohan Garden
NEW DELHI 110059 (India)

Lasertypeset by
Microtech Advance Printing Systems (Pvt.) Ltd.
H-13, Bali Nagar
NEW DELHI 110015

GENERAL EDITOR'S INTRODUCTION

Environmental Geography is progressively emerging as an independent discipline itself. Experts may differ in the basic concepts, definitions and approaches in encompassing various aspects of environmental geography. But the fact is that, this is an inter-disciplinary subject and various parameters of Physical Geography suit well to explain main linkages and intricacies of environmental changes in relation to land degradation, flooding, erosion and natural hazards on a firm scientific foundation. It is so because the environmental geography is mainly concerned to provide a rational interpretation of various variable environmental parameters over space.

The author, Dr. A.A. Pirazizy, has chosen certain well established geographical parameters to study the natural hazards in an inter-disciplinary manner including the physical, bio-physical anthropogenic systems, and also by assessing natural hazard as erosion and has suggested suitable models to mitigate the damages. The research design criteria adopted and the problems likely to be faced in such studies of environmental geography, are well documented.

I hope, the critical evaluation of the problem and risk assessment analysis presented, will create an urge amongst researchers and students for adopting an inter-disciplinary integrated research approach to understand environmental geography and present suitable models for environment management programmes.

Breakthrough Communication Centre
New Delhi 110015
15 June 1992

K.V. PALIWAL

PREFACE

The ineptitude of the anthropogenic conception regarding environmental anomaly, least perhaps the researcher, to interact with major, yet infrequent discontinuities resulting from gradual, imperciptible modifications sustained for long periods; especially in case of highland environmental interactive systems.

The present book is anticipated to trigger the interest of various disciplines, but specifically it will be helpful to college students of geography and environmental studies. The main stress has been to highlight the demurs that are still engulfed in incertitude.

Environmental geography does not pursue rudimentary explanations for environmental demurs within any established discipline, but rather in the interstices and linkages between disciplines. Thus, environmental geography is a demonstration of the fact that the major changes in the approach in the environmental matters are inexorable. Environmental geography, though not yet an established descipline in geography, is a dynamic progression of physical geography, attempting a synthesis of the various interrelationships that are responsible for life, elevation, slope, relief, quantity and quality of degradation and agradation, crustal disturbances, process of orogeny, climatic variations, and patterns and distribution of flora and fauna.

As environmental studies have been awed by a multiplicity of facts and theories, treating issues that are peripheral to the main concern, the environmental geography endeavours are comprehensive by concentrating on facts, and central principle issues that intimidate the natural environment. The environmental discipline of geography thus, treats elements of environmental demurs as systematic phenomena, and not separate as physical, biological, social cultural, and historical phenomena. These and other elements of the environment are inextricably linked in order to produce a physical world, and which are the main theme of scientific consideration of environmental geography.

The main objective of the present book is to indicate the character, and approximate, the extent of changes caused in physical conditions of environment. It further intends to depicting the damages of imprudence, and the necessity of caution in all operations which on large scale, interfere with the spontaneous arrangements of the environment. First step in such attempts, thus, would commence from constructing a research design which attempts to test the assumptions that are presently established themselves as irrefutable. Thus, such a major problem becomes extremely difficult to manage, unless an integrated interdisciplinary approach is adopted.

A.A. PIRAZIZY

CONTENTS

SECTION III
Hazards and Risks

LIST OF TABLES

LIST OF FIGURES

LIST OF PLATES

SECTION I

AN ENVIRONMENTAL APPRAISAL

"Ah! is it that those Himalayan solitudes, cooled by the spray of minute bits of Ganga's waves and abounding in beautiful rocky flats such as are the haunts of Vidyadharas*, are all engulfed in destruction, that men in disgrace hang on others for their maintenance?. Or is it that herbs and roots have all disappeared from caves, and streams have gone away from hillsides, or that branches of trees bearing luscious fruits and yielding barks are all destroyed, that the faces of wretches, perfectly devoid of good breeding, are found to have their eyebrows dancing like creepers in the wind of an arrogance which their scanty earning eked out with hardship engenders in them."

— *Vairagya-Satakam*, 24,25

---

* Vidyadharas are unearthly beings with super human skill in arts, especially music.

CHAPTER ONE

AN OVERVIEW OF ENVIRONMENTAL IMPACT ASSESSMENT

1.1 **Prelude**

The present book is an approach of environmental geography, aiming to study hazards and appraise the ways that the eventual risks can be reduced, if all damages cannot be mitigated. The scientific aspects of geography include study of the physical components of environment (Deorr, 1990). Environmental geography, though not yet an established discipline in geography, is a dynamic progression of physical geography, attempting a synthesis of the various interrelationships that are responsible for life, elevation, slope, relief, quantity and quality of degradation and agradation, crustal disturbances, process of orogeny, climatic variations, and patterns and distribution of flora and fauna. These and other elements of the environment are inextricably linked in order to produce a physical world, and which are the main theme of scientific consideration of environmental geography (Pirazizy, 1991). Though these linkages are infinitely intricate, yet scientific approach of environmental geography with its interdisciplinary organization is slowly probing associations, thus, developing theories and models to serve as platform for prognostications.

Environmental geography does not pursue rudimentary explanations for environmental demurs within any established discipline, but rather in the interstices and linkages between disciplines. Thus, environmental geography is a demonstration of the fact that the major changes in the approach in the environmental matters are inexorable.

An environmental studies have been awed by a multiplicity of facts and theories, some of which contradicting others, and as most of environmental literature treats issues that are peripheral to the main concern, the environmental geography endeavors are comprehensive by concentrating on facts, and central principle issues that intimidate the natural environment. The environmental discipline of geography thus, treats elements of environmental demurs as systematic phenomena, and not separate as physical, biological, social cultural, and historical phenomena.

The present book is anticipated to trigger the interest of various disciplines, but specifically it will be helpful to college students of geography and environmental studies. The main stress has been to highlight the demurs that are still engulfed in incertitude. The treatment that environmental geography proposes towards the assessment of hazards and eventual risks, is elementary and descriptive, rather than mathematical in real sense. Nevertheless, the approach is analytical and critical, highlighting the controversy and dissimilarities of view points in dealing with the problems of hazards. Environmental geography proposes a candid approach to those interested in mitigating environmental problems. The approach advocates to directly and solidly encounter the complexity of the natural environment along with economic forces, and the deficiency of diligent and complaisant theoretical models.

Thus, within limits, the anthropo-natural (environmental) dichotomy, on environmental arena of interactive systems, is the commencing point. Natural pressures being with the geomorphological forces, are the causes of endogenic and exogenic mountain building and alteration forces. The presently active process of orogeny in Himalaya, combined with precipitation and action of runoff is steepening the slopes, hence, eventual instability, which is considered to be hazardous in pretext of anthropogenic ventures and occupancy. The natural hazard problems associated with highland Himalaya are natural processes having a context of steep slopes and high relief energy that makes gravity related problems, such as avalanches, slides, slips and runoff, more stern. The essential conundrum being that the modern developmental strategy of these highly fragile environments has tended to apply lowland technology to settings scarcely congenial. Yet, it is difficult, if not impossible (Cool, 1983), to quantify the actual proportion of the total pressure that can be attributed to nature, as contrasted to anthropogenic forces. Thus, explanation of pressures, and their impact on highland environment are not confined to the limited sphere of cause and effect. These are complex interactive systems, each possessing an internal dynamic cohesiveness; each interacting with and impinged upon by a myriad of other natural and anthropogenic systems. These systems are always adjusting and modifying themselves in relation to each other. Thus equilibrium of highlands must be seen in dynamic terms (Pirazizy, 1989).

Environment is a dynamic open system, always in the process of change, irrespective of place of prevalence. There are two ways of approaching the environmental problems of the Himalaya. Either to seek the facts or to seek what would the scholars and concerned groups want the facts to be. Here most of the emphasis is being laid upon the second approach through an explanation of the vast uncertainties that are a marked feature of the zone. Long term changes in environment through geological evolution or/and short term modifications such as climatic fluctuations or/and the hazardous changes such as earthquakes, mass movement and the like, have always threatened anthropogenic survival and adaptation in highlands. While dealing with Himalaya, intrinsic dynamism of these highlands must be signified. This dynamism is consequent to rates of orogeny and regional denudation in context of geologic time scale. Doctrine of plate tectonics upholds the dynamic process of cataclysm of Himalaya. This considerable scale of upheaval leads to leviathan mass erosion.[1] This phenomenal erosion and regional denudation, which generates huge sediment transfer and deposition, has been vigorous over the past million years.[2] These rates of uplift are presently a great deal prominent than that of denudation, thus, making the Himalaya an extremely dynamic realm.

The aforementioned observations are affirmations of the fact that highland environments prevail and strive in severe conditions, apropos anthropogenic perspective. The physical environment, comprising of soil cover, flora and fauna have evolved in absence of man, meaning that either the impact of man on their rate of evolution had been nil, or processes of their evolution dates much beyond the appearance of human species on the scene. Thus, they are highly susceptible to the presence of human beings. This vulnerability actuates a great deal of irreconcibility between utilization and environmental integrity maintenance. The most active feed back in the technocratic phase of anthropogenic interaction has been morphogenetic, which in no way leads to original equilibrium or any equilibrium, rather it characterizes an environmental system with a certain relationships between altered natural constituents, either through natural or anthropogenic impacts or both, to be modified into another system with different relationships, characterized by lesser ability of its components to support biological life.

Highlands are dynamic environments staging both autogenic and allogenic forces. These forces lead to drastic transformation of the highlands as far as their overall environments are concerned. This natural instability as a result of incessant upheaval and denudation, however, is accelerated and energetically spreading. Though, on regional scale highland anthropo-environmental interactive systems are perceived to be of no critical impact at present, as Ives et al. (1987) remarked:

> ". . . it would be reasonable to argue that the contribution to human interventions over the past three or four decades, or even centuries, has been insignificant when balanced against the natural processes at work."

Yet, on microscale, these interactive systems, can/have faced rigorous change with a strong shift towards deterioration, apparently irreversible for near posterity. The consequent of this impact is believed to be disturbing the lowlands, which are recipients of displaced material and migrants. Eventhough there is an consciousness regarding such types of environmental deterioration, the generic causative factors are equivocal.[3] The cognition of these systems has been a pressing challenge, because of the enigmatic environmental conditions, and rapidity and scale of deterioration which restrains the assessment of modifications and resulting eventualities. A broad alignment of these environmental systems, describes them as being topographically extremely undulating, having poor soil cover with marginal availability of nutrients, uncertain precipitation, marked variations in diurnal temperatures, frost occurrence, and low temperature. Natural unpredictability, low carrying capacity, high ecological fragility, and rigorous down slope movement of material are auxiliary detrimental factors, impeding anthropo-environmental interactions in these environments.

A general uncertainty regarding the unequivocal impacts of anthropogenic enterprises upon the highland environments is the present phase of environmental research (Pirazizy, 1991). This uncertainty emphasizes that there ought to be a distinguish between what is natural and what is accelerated through the anthropogenic activities. This distinction, must not be based on the content and nature of other's experiences but on the way in which they are controlled and on the degree to which a researcher is able to integrate these impressions. Nevertheless, the ability to encounter and utilize the environment, on ecologically concurrent basis, depends on this research and procedure of its portrayal. Despite the dominant uncertainty, both mismanagement and ill planning in anthropogenic activities, such as, production, construction, recreation, and like, have resulted in further instability through erosion and degradation hazards such as damages to productive lands, construction sites, natural and anthropogenic drainage systems, subsurface water flow, and impoverishment of water resources, which in turn detrimentally affect socio-economic conditions of inhabitants. Examples such as road, hydal-project, and building construction, through utilization of highly inclined surfaces subsequent to ill-planning, are many.[4]

1.2 **Publications**

Environmental studies in their present perspectives are of contemporary conception. The modern millennium of interest in environmental perception in geography commenced from the paper published by Wright (1947). Nevertheless, the first systematic study regarding the anthropogenic perception of the environment was accomplished by Trowbridge (1913). Yet 1960's marked the large scale inquisitiveness in environmental knowing. This period marked the promotion of research lines in geography and anthropology. In geography anthropo-environmental relationship was conjured as a part of behavioural geography. Later on research lines were dictated by the works of White (1945), Kates (1963, 1970), Burton and Kates (1964), on the cognizance of natural hazards, Lowenthal (1968), Lowenthal and Prince (1964,1965), on the perception of landscape, Wolpert (1964), Gould (1963), Golledge (1967), Golledge and Moore (1973), on behaviour and spatial preferences.

In 1970s lines of inquiry moved to geosophy and ethno-geography. Geosophy, in other words the transforming cognition of the environment as seen over historical time was perused by Bowden (1971), Allen (1972), Jackson (1970), and the others. Ethno-geography, or the conception of people of dissimilar cultures regarding the environment was solicited by the works of Knight (1971).

Specifically speaking the study of environmental issues in highlands subsequent to anthropogenic impacts, is of much recent injunction. The weak geological configuration of Himalaya, being a dynamic stage of strong autogenic and allogenic coercions of change, has played a considerable role in luring the attention of scholars and scientists both inter and intra nationally in 1970's. This interest in Himalaya and its environment was, grieving, not congruous to all of these elevated land masses. Himachal Pradesh, for instance, unlike its neighbouring highlands of Nepal, Kumaon, and Gharwal Himalaya, has not received adequate attention.

The interest in Himalaya has a historical connotation. Impour of travellers into Himalaya is marked distinction of 19th century. These travellers were interested in rituals, traditions, aesthetic vistas, enchanting natural environs of Himalaya. The vanguard treatises on Himachal Pradesh started by the state's inclusion to British administration. Basic issues then were those concerning land management for revenue, grazing problems caused by transhumants, forest deterioration, and the like, which were produced in works of Duthie (1883), Brandic (1876), Lyall (1875). Works on Himachal Pradesh then occurred in publications of Anderson (1898), Drew (1877), and Aitkinson (1886), based on problems associated with forestry, grazing, and related environmental issues as they discerned it.

During 1900–1930, works of Trevor (1920), Ibbetson (1914), Holland (1928), Conventry (1929), and Hutchinson and Vogel (1933), dealt extensively with social niche, forestry, grazing, ecology, and malevolent erosion conundrums of the zone, according to their dexterity. Documentations on account of journey, scenic pulchritude, geology, biology, history, settlement, wild life, and geography of Himalaya during the same period, in general, is accessible in the works of Collet (1902), Back (1925), Jones (1905), Beresford (1925), and Whistler (1924). From 1930 to 1985 extensive literature was generated on Himalayas in general dealing with various scholastic facets of the environment.

The literature that covers the range of knowledge required for the present study is burgeoning. Academic and applied studies of high altitude problems have flourished for the last few decades. Among the most informative of recent texts are *Landslide: Causes, Consequences and Environment* by Crozier (1986); *Hill slope Stability and Landuse* by Sidle et al (1985); *Landslide: Hazard Zonation* by Varnes (1984); *Landslide Analysis and Control* by Schuster and Krizek (1985); *Process in Geomorphology* by Embelto and Thornes (1979); *Landslides and Their Control* by Zaruba and Mencl (1969); *Geomorphology in Environmental Management* by Cooke and Doornkamp

(1990); *Geological Hazards* by Bolt et al (1975); *Disaster and Reconstruction* by Geipel (1982). Much relevant research is published in generally available journals, the most useful of which include, *Mountain Research and Development, Geojournal, Geoforum, Ambio, American Association of Geographers, Earth Sciences, Canadian Journal of Earth Sciences, Canadian Geotechnical Journal,* and *Progress in Physical Geography*, and some Government reports and memoirs such as *Indo-German Reports, Himachal Pradesh Environmental Status Report, Himachal Pradesh Grazing Advisory Committee Report,* and *Bulletin of River Valley Ministry of Agriculture, Government of India.*

The theory of mountain environment degradation was gradually consolidated from proposals of various scientists, academicians, and various research scholars working in these lines. This theory is based on the scenarios of irreversible degradation and imminent calamity. This theory was indirectly and vernacularly refuted by ICIMOD, Dec., 1983 session, as mostly being repetitions of what had been the scheme of 1974 Munich conference. With this aphorism the session set a procedure, thus, subjecting the presented themes during this period to verification regarding their validity. However, Ives (1985), highlighted this aspect in the theme of *The Theory of Himalayan-Indogangetic Plains Environmental Degradation*, as being widely accepted theory, which in part is mythical, having been enriched by years of *emotional* repetition of first impressions.

1.3 **Factual Environmental Status**

There are various debates regarding the environment, each purporting concern for the state of both natural and anthropogenic environments. There is generally a widely accepted agreement dealing with the fact that anthropogenic interaction and relation towards natural environment, in the anthropo-environmental interactive systems must improve. This great and widespread concern about the plethora of issues related to environment are characteristic occurrence in academic research, and all these depend on environmental perception, as Carr (1967) expressed the view that:

> "We ask why we should be concerned with the form of the environment and how . . . [it] might contribute to the quality of life . . . [the reason being that] we lack much relevant knowledge,

> but worse . . . of getting beyond our current [traditional] conceptualization of . . . [environment] form and its functions. Indeed, the most perplexing problem we face in attempting to improve the relationship of people with their . . . environment is the persistence of conventional images and models of conceiving the . . . [environment] in the face of changing . . . [environmental] realities and human purposes."

Environmental perception is defined as the sum of images, information, and impressions (Moore and Gollege, 1976), people have regarding the modules, conformations, and performances of physiocultural socio-cultural, socioeconomic, and sociopolitical politico-economic environments. These perceptions, however, are treated as subsets and as a function of environmental studies.

The perception affect the human relation in anthropo-environmental interactive systems. Anthropogenic relation to nature is mostly based on the idea of nature in its physical reality. These peculiar dispositions of living with perception determine the set of interactive parameters and are of prime importance in the appraisal of anthropogenic impact on environment. It may be said that the holistic approach to environment is not achievable unless a unity of perceptions based on the real state of environment is not reached as stated by Bowman (1934):

> ". . . the natural environment is always a different thing to different . . . [groups and changing in accordance with the changing priorities]. Its [natural environment's] potentialities are absolute, but their realization is a relative matter, relative to what a particular . . . [group] wants and what . . . [they] can get with the instruments of power and the ideas at . . . [their] command and the standards of living . . . [they] demand . . . to attain."

Notwithstanding, these perceptions are of two characteristics, either *Figurative* (Piaget, 1967), emerging from contiguous association, and/or operative, pertaining to anthropo environmental interactive systems. Thus, depending on the procedure of investigation, reconnaissance, and comprehension, progression of perceptions take place as stated by Moore and Golledge (1976):

> "No one person's or group's image of environment is more

"accurate" or true than another's. Each views and reconstructs the world through its own assumptions, values, and life experiences."

The sources of confusion related to environmental research, especially those confronted by environmental geographers are as follows:

- Long sequences of components connected by cause and effect, not being apparent to most observers regarding origination of causes and termination of effects;
- Misunderstanding resulted by deficiency of information, suppression of correct information, specious information; and
- Imperceptibility of slow and aberrantly changing subtle environmental phenomena.

The ineptitude of the anthropogenic conception regarding environmental anomaly, least perhaps the researcher, to interact with major, yet infrequent discontinuities resulting from gradual, imperceptible modifications sustained for long periods; especially in case of highland environmental interactive systems.

Environment of highlands in general and Himalaya in particular are ecologically fragile in nature. In highlands with the increase in altitude, the capability to attune environmental disturbances declines. Himalaya being young mountain systems, which are currently active in the process of orogeny, comprise of geologically tenuous formations (excepting a few tracts), steep slopes great range of altitude, and uncertain climatic conditions, are immensely vulnerable to disturbances. One of the major factors of these disturbances are mismanagement, ill planning, and malexecution in the present phase of anthropogenic ventures. If this level and scale of interaction continues, the aftermath will be an undesirable environment which could be the only resource base for sustenance of thousands of dwellers, who are equipped with a technology which cannot compete with the state of transformation. Thus, the basic factors in mountain research are highlands, lowlands, and slopes in between, which occupy the greater area and are the causal stages in appended geomorphic processes, as remarked by Lynch (1960):

"A clear image [of environment] enables one to move about easily and quickly. . . . But an ordered environment can do more . . . it may serve as . . . an organizer of activity or belief or knowledge . . . play[ing] a social role as well."

The representations in the present book are treated as hypothetical contrast, which proposes about processes and eventualities not directly discernible, extending beyond observed data. The representations have been attributed a real status, referring to postulated underlying entities or processes that there is a reason to believe do exist, meaning observable in principle, but presently capable of being hypothesized.

Degradation of Himalaya, however, is a supposed conjecture of various landuse patterns, for instance, contemporary deforestation, overgrazing, wasteful agricultural practices, mismanaged developmental ventures, which are held responsible for environmental degradation. Presently data regarding the natural rates of regeneration in the zone is not available, which could have reduced the severity of appraisal of human induced rates. As stated earlier, there is little doubt about the evolution of highland ecosystems, their biota and soils, in absence of human beings. The highland soils, the life support system source, are shallow and formed on steep slopes, and protected by the native vegetation cover. These soils are highly susceptible to erosion if this protective mantle is impaired. Highland soils in order of erodibility can be categorized as silt, silt loam, loamy sand, sandy loam, and loam. These soil types, however, are the basic soil types of the highlands. Ironically, most of the productive soils in highlands are comprised of fine particles such as, silt which is a fine particle of sand, highly prone to erosion. One of the major human correlates, i.e. agricultural activities, that are operated through multifarious modes, disturbs this equilibrium rate, by replacing the natural vegetation cover with the domesticated ones, hence, the accelerated erosion with subsequent negative impacts on highland environment. Meanwhile, another anthropogenic activity, i.e. terracing is a method of reclamation of erosion affected lands.[5]

Increase in population, creation of various occupations, means of livelihood, demands, need for more productive lands, being estimated as one of the major causes in the process of deforestation, along with making steep slopes bare of adequate protective cover, being referred to as causative factor in increased soil erosion, generation of landslides and mass movement, and disruption of normal hydrologic cycle, and increased runoff during rains, enhancing factor of hazards, are considered as major causes of highland environment deterioration in the scenario of Himalayan-Indogangetic Plains Degradation Theory. However, basic questions are that:

(i) These assumptions are supported with data which has been claimed to be unreliable in many aspects, having penetrated into massive literature as a standard media for perception and research works;[6]

(ii) The holistic cycle of nature is reduced to fragments and entities which are but one in the chain of events, yet analyzed separately. Bateson (1972) stated that:

"logic can be used in very elegant ways to describe linear systems of cause and effect, but when causal sequences become circular, as they do in the living world, their description in terms of logic will generate paradoxes."

As remarked by Chew (1985):

"Nature cannot be reduced to fundamental entities, but has to be understood entirely through self consistency . . . [nature's] components . . . [are]consistent with one another and with themselves."

Thus, such a major problem becomes extremely difficult to manage, unless an integrated interdisciplinary approach is adopted. The integrated development of highlands is the procedure of resource use which ensures a sustainable utilization of these resources with the means of present technical knowledge for a few generations ahead. Schumacher (1975) indicated:

". . . [environmental problems are] remnant of nineteenth-century thinking . . . [which] is . . . incapable of solving any of the real problems of today's environmental degradation, due to lack of holistic interdisciplinary approach] . . . [The attitude] is fragmentary and reductionistic, restricting itself to purely quantitative analysis and refusing to look into the real nature of things . . . [This is the deficiency of] the modern technology."

First step in such attempts, thus, would commence from constructing a research design which attempts to test the assumptions that are presently established themselves as irrefutable (Pirazizy, 1991). The purpose of the present book, thus, perpetuates in to a collective summation of following points in order to ascertain:

(i) The actual modes of modification, degradation, and destruction in the form of erosion and hazards of natural environments in temperate highlands, with regard to anthropogenic impact?;
(ii) The actual causes of erosion and hazards and the role they play in highland-lowland anthropo-environmental interactive systems?;
(iii) The actual rates of environmental anomaly in different elevation/ vegetation cover/aspect in the zone?.

The main objective of the present book is to indicate the character, and approximate, the extent of changes caused in physical conditions of environment. It further intends to depict the damages of imprudence and the necessity of caution in all operations which on large scale interfere with the spontaneous arrangements of the environment. It means to suggest the possibility and the importance of the retention of disturbed harmonies in the exhausted region.

1.4 Structure of Assessment Process

Decisions on resource utilization and environmental protection must be based on detailed and reliable environmental data. Knowledge of landscape and landuse modifications has become an increasingly important prerequisite for better use of the environment. Environment must be carefully and constantly monitored to avoid hazards associated with uncontrolled development and deteriorating environmental quality. Thus, the structure of assessment process in the present book gives definitive attention to following environmental situations and problems in the temperate zone which are as follows:

- Maintenance of environmental quality;
- Prevention and reduction of anthropologically accelerated soil erosion;
- Recognition of probable hazards;
- Maintenance of fragile quality and integrity of natural environment against some types of developmental ventures; and

- Mitigation of transportation corridor impact on environmental conditions.

A landscape is an empirical evidence depicting the interaction of man with his environment. Landscape perception and evolution is depicted in its two specific functions, *Hazard Evasion*, and *Directive on Suitability for Dominant Landuse Category*. Evaluation of landscape is an important aspect in any study related to environmental issues and considerations, due to their potentiality to enhance resource litany, carrying capacity decisions, and environmental impact appraisal. Evidences on the evolution of landscape in any region, through interaction of diverse physical parameters, such as configuration, perspective, vegetation, climate, and like, and anthropogenic endeavour paradigms depict the inherent Ecosystem Inertia (abstain to transform), and Ecosystem Resilience (capability to ameliorate from disturbances).

The prevailing presumptions are scrutinized empirically by making logical inferences from them. The justification of the conjectures, eventhough, not a sufficient prerequisite to prove the vitality of the assumptions, is imperative, endowing substantiation for their vitality. They, however, are convenient assumptions for a set of processes beyond the practical confines or immediate inquisitiveness of researcher, but which nevertheless, are useful in accounting for other observable spatial processes.

Many situations or conditions exist in which a proper balance must soon be struck, if further anomaly, generally unacceptable to anthropogenic safety and maintenance of environmental quality and integrity, is to be avoided. Even more crucial is the necessity to recognize the ever increasing risk and heavy restoration costs that will inevitably result from continued disturbance through anthropogenic ventures. Thus, the knowledge of landscape can provide useful and rudimentary information for the surveying and monitoring of environmental situations and problems.

However, a well organized standardize approach to the collection and presentation of landscape information should provide an essential element in any system for monitoring of environmental conditions.

In an attempt to peruse and determine the anthropogenic impact on environmental transformation in temperate zone of Himalaya, the State of Himachal Pradesh has been selected (Fig. 1.1). This state on the basis of natural vegetation cover, climate, soil, and altitude has been divided in to three zones, i.e. Alpine, Temperate, and Tropical. Out of these three zones the temperate zone, which in itself further subdivided into moist and wet temperate sub zones, is selected as the study area. Selection of this zone can be explained on the following considerations.

- Most of the anthropogenic activities on highlands are located in temperate zone; and
- Temperate zone is also considered to be the ultimate belt in highland environment in which adaptive commencement and retention of anthropogenic ventures regarding land utilization and survival is possible.

If the Himalaya are implied to as macro-level region in sequence of highland-lowland environmental investigations, the state of Himachal Pradesh was assigned the meso-level characteristics for research and study, thus, the area constituting the empirical evidence can be accredited of being micro-level. The plan of probing however, subjected two districts of Kullu, Chamba, and Shimla to in depth appraisal for hazard and risk, and erosion, sediment yield, and mass wasting assessment and their interrelationships with each other and anthropogenic ventures and natural rates as highlighted in preceding sections. The appraisal in general can be condescend to probe erosion, suitability, capability, natural hazards, and risk classification.

To study the anthropogenic impact, various anthropogenic ventures such as road, highway, and hydal project and building constructions, falling under developmental activities were considered. Deduction of environmental processes is presently an epistemological issue which can be governed through system theory. System theory asserts that behaviour and knowledge are strictly under the control of environment. Data collected, however, in themselves do not combine into a composite picture of the environment. The relations among the data from various sources are established through habits, costumes, and the laws of association as expressed by Ittelson and Cantril (1960):

> "Perceiving is that part of the process of living [environment] by which each one of us, from his own particular point of view, creates for himself the [environment] in which he has his life

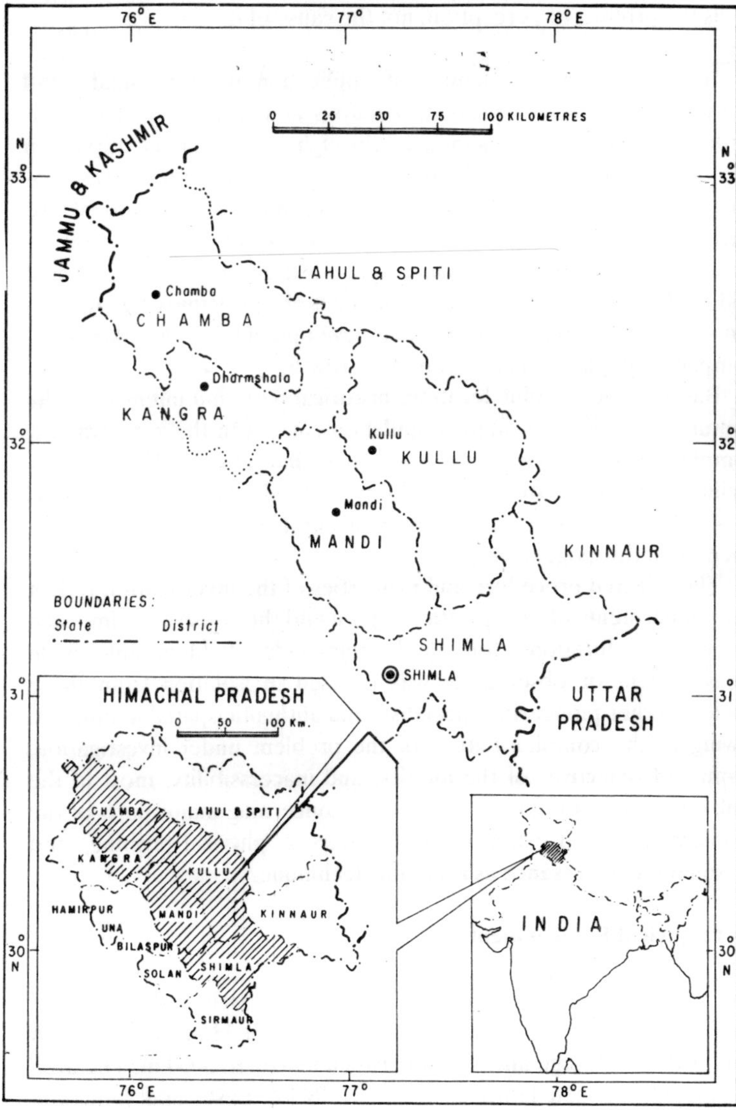

FIG. 1.1: Temperate Zone of Himachal Pradesh

experiences. . . . Independent of experience, we can nevertheless assert that the . . . [environmental issues have] no meaning and cannot be defined. . . . The [environment] . . . as we experience it is the product of perception, not the cause of it."

In addition their, manner of application were pursued, and spatio-temporal status probed. Analysis and deduction of the environmental modifications induced through anthropogenic impact, in the zone within the limited range of natural environment, was conducted applying various computation programmes. Subsequent impact assessment of highland temperate anthropo-environmental interactive system through measurement of variations in sediment yield and load along the main gauges situated in the principal drainage systems, and pace and progression of developmental undertakings accompanied population and economic growth.

Based on geographic location, historical past, and intensity of the demur, a selection of major population centers in the zone was accomplished to investigate the socio-economic effects. This scrutiny facilitated the evaluation of capability, suitability, and stability areas which stage the environmentally unfavourable modifications subsequent to anthropogenic impact.

The selected procedure and properties of the investigation reflect the requirements of the specific purpose and the constraints imposed by impost, duration, toil, and promenade. Seldom any single measurement or observation was perused in isolation from those made by other pioneering specialists and authentic spatial anomalies. Owing to the complex nature of the problem under investigation, demurs of objectives of the inquiry, and inaccessibility, most of the time deficiency of infallible inventories concerning major parts of the problem, impelled extensive field traverses, physical probing, and personal reconnaissance as a suitable technique.[7]

1.5 Empirical Environment

The area, selected to supplement the approach with empirical evidences, is located in the temperate zone of Himachal Pradesh state with an average altitude of between 900 m. to 5000 m. The area is situated between latitude 75°52" to 78°19" East, and longitude 31°44' to 33°13' North, covering an approximate area of 20,000 km^2. About 25 tehsils constitute the research area. These tehsils have been selected from five districts out of twelve districts constituting the state.

From Chamba district, four tehsils of Chaurah, Battiyat, Chamba, and Brahmaur have been selected. Whole of Mandi district with its six tehsils and four sub-tehsils were chosen. The six tehsils and four sub-tehsils of the district are, Joginder Nagar tehsil, Lad Bharol sub-tehsil, Sandhol sub-tehsil, Sarkaghat tehsil, Mandi tehsil, Sunder nagar tehsil, Chachyot sub-tehsil, Chachyot tehsil, Bali Chock sub-tehsil, Karsog tehsil. From Kangra district contributed only two tehsils of Kangra and Palampur were incorporated. Main emphasis was concentrated on entire Kullu district with its four constituent tehsils, i.e. Kullu, Banjar, Ani, and Nermand and Shimla district along with its eight tehsils and three sub-tehsils, i.e. Seoni, Shimla, Theog, Rampur, Ruhru, Jubbal, Kotkhai, Chaupal tehsils, Nankhari, Kumharsain, and Nerua sub-tehsils.

The inquiry perimeter is located in a highly undulating mountainous region with an average elevation of about 1800 m. above mean sea level. A condensed history of the zone is vital in discrimination of its perspectives of development in context of the modern technological era. Major changes in administrative boundaries took place during 1961–71 decade. On account of "The Organization of Punjab State Act, 1966", districts of Shimla, Kangra, and Kullu were added to the State, as a result the number of districts in the present delineated temperate zone, increased to five in 1966 against the two in 1961. Besides, three towns of Dalhousie M.C., Dalhousie C.B., and Bakloh C.B. were included in Bhattiyat tehsil of Chamba district from the Gurdaspur district of Punjab.

After 1961, Kullu tehsil was given a district status and got separated from Kangra district. Later on three tehsils were taken out of the Kullu district, presently known as Banjar, Ani, and Nirmand. In Mandi district, 23 villages of Joginder Nagar tehsil, and one of Chichyot tehsil were incorporated in Mandi Sadar tehsil. In 1966, 231 villages of Kandaghat tehsil were contained in Shimla sub-tehsil, and Nalagarh tehsil was sculpted out of Ambala district into Shimla district. During 1971–81, Kasumpti tehsil was abrogated and associated with Shimla sub-tehsil. Thus, Shimla district was structured by coalesces of tehsils and sub-tehsils of, Seoni, Kumharsain, Rampur, Rohru, Jubbal, Kotkhai, Theog, Kasumpti, Chaupal of Mahasu district, and Shimla sub-tehsil of old Shimla district. In Chamba district, Saluni and Sihunta tehsils were fashioned out of Chaurah and Bhattiyat tehsils respectively, together comprising of 351 villages. In 1980, four unprecedented sub-tehsils of Lad Bharol, Sandhol, Bali Chowki, and Chachyot were conceived in Mandi district. In Kullu district three villages were transferred from Banjar to Kullu tehsil. In Shimla at the same period, Nerva and Nankhari sub-tehsils were formed out of Chaupal and Rampur tehsils respectively.

1.6 Organization of Processes

It was against the above background of shift in emphasis within the environmental geography, that forms the basic approach of this book, that the idea to compose the present volume was conceived. It seems pertinent to envisage a volume specially written to analyse the significance of anthropogenic ventures to the range of environmental processes, studied by the environmental geographers, the forerunners of a new discipline which needs acknowledgement and place of its own in geography as a bridge between various natural and social science disciplines. However, the contents although composed in the light of recent foci in environmental geography should be of interest to other disciplines too.

This volume is directed primarily towards highland-lowland environmental interactive system processes with chapters grouped under major headings discussed in proceeding paragraphs. Some spheres of the sections were given less attention, partly inevitably and partly as a reflection of research activity of environmental geographers and their tendency to ignore some sections. The alluded sections have been distinctly chapterized as follows:

SECTION I: An Overview of Environmental Impact Assessment

This section is conceptual, dealing with the demurs, factual processes, structure of investigation, delineation of empirical environment, and a perception analysis. This section is comprised of only chapter one.

SECTION II: Biophysical Systems

This section, consisting of two chapters i.e. chapter two and three, deals with interdependence of physical components of environment. In this section natural environment is scrutinized through evaluation of geology, slope, soil, relief, geomorphologically related items, climate, vegetation and drainage. The principle objective has emphasized on the appraisal and cognition of landscape in environmental perspective.

SECTION III: **Hazard Assessment and Anthropogenic Endeavours**

This section is comprised of two chapters. Chapter four anthropogenic endeavours, deals with major developmental activities. The objective of this section emphasizes on the assessment and cognizance of multifarious anthropogenic activities. The utility patterns, categories, situations, contemporary trends, and environmental impacts regarding the construction ventures compose the inherent subject matter of this section. The structure of assessment in this section has stressed on delving into environmental demurs created by both ill planning and mismanaged developmental ventures, and inherent environmental vulnerability of highland systems. and Chapter five natural hazards and eventual risks appraisal, deals with definition and differentiation in between natural and accelerated hazards. This section mainly emphasizes on the occurrences, modes, frequencies, magnitudes, mapping, and prognostication of future scenarios associated with natural hazards.

SECTION IV: **Eventualities of Environmental Anomaly**

In this section erosion appraisal is the main theme. The fundamental significance is ascribed to the problems of erosion, its measurement and modeling, suitability and sustainability, and subsequent realities, which have been discussed and in profoundity annotated.

SECTION V: **Prognostic Postlude**

This section is an overview of the book concentrating on anthropo-environmental interactive system in highland temperate Himalaya. Cases of environmental anomaly are elaborated and conclusions derived. Policy connotations have not been given a separate section, rather are embedded through out the book in appropriate places. This section in aggregate varifies the assumption of highland anthropo-environmental interactive system, substantiating its assertions by corroborating evidences attained in preceding sections.

NOTES

1. Sharma (1983) stated that, drill cavities have perforated more than 5000 m. in alluvial sediments underneath the gangetic plain.
2. Geophysical inferences indicate that the height of Himalaya is not less than what it was a million years ago. The rates of uplift for Himalaya have been proximated by different scientists as 1 mm/yr for Great Himalaya, about 9 mm/yr around Nanga Parbat region (Zeilter, 1982), and 1–4 mm/yr since the close of Pleistocene (Low, 1968).
3. Information regarding these problems is fragmented and most of the time not available, rendering it inadequate data base for empirical investigation of the issues relating to highland anthropo-environmental interactive systems. Most of the works produced regarding the highland environmental degradation are literary with sentimental petitions, asserting the impression that the Himalaya are facing destruction. The warning has acquired such a dimension that the belief of destruction of most of Himalaya by the end of century or at most the early next century is a general expectation.
4. For instance, according to the BACON, about 550 m^3 of debris per kilometer of roads constructed in mountainous tract of western Himalaya is produced annually through landslides and rockfalls, causing about 24 x106m^3 of sediment slide down the slopes. The above factors all lead to a decline in life quality with sequence of vicious effects.
5. Ives (1979), while visiting Kakani area, Nepal, in initial response to the large amount of landsliding and soil erosion, comparable to appraisals of the earlier works, stated that this area was at the periphery of calamity. But during his later visits, he discovered that wet and dry agricultural terraces, traditional village level reclamation techniques, had tended these defected lands of the area. He stated that, if those efforts were deduced from the land loss assessments of the single visits (customarily), the rate of environmental degradation on such a colossal scale, would be reduced to much lesser quantities than initial projections.
6. As an example Ives and Messerli (1989), reveal that, existing facts on per capita fuel wood consumption in Himalaya vary by a factor of 27, which impedes any rational appraisal and planning action with such a wide variation. Moreover, the fuel wood consumption, increasing population, growing needs, and environmental deterioration are the basis or conclusion of most of the studies from which this factoral variation of 27 is calculated.
7. As an example following instances are able to illustrate the need for personal reconnaissances and the trend of their conduct for variegated purposes. While construction of natural hazards risk assessment map, kilometres long stretches of Kullu-Shimla, Mandi-Shimla, and Rampur-Kinnaur highways were traversed. Besides, for assessment of mentioned problems in interiors and anthropogenic centres, Shimla-Fagu-Kufri-Theog-Narkanda stretch, most of Pabar and Giri river courses, Shimla-Bilaspur stretches too were subjected to the same probing. While appraising landslides and analyzing suitability measures for enacting large scale construction activities, the worst affected stretches of Mandi-Manali, Chamba-Tissa, and Chamba-Brahmaur roads were travelled on foot. On account of assessing soil erosion through erosion peculiarity and silt yield index, in collaboration with PWDHP engineers and a

private organization estimating environmental pollution (based in Lucknow), about 15 sub-catchments of Sutlej, Giri, and Upper Beas catchments were appraised. With regards to major part of the data, forming basic tool of physical analysis, the modified outline of the scheme fostered by the Ministry of Agriculture (1972), was adopted. On the basis of this procedure primary reconnaissance of the units selected through sampling method considering the procedure of the study.

SECTION II

BIOPHYSICAL EQUILIBRIUM SYSTEMS

BIOPHYSICAL EQUILIBRIUM SYSTEMS

The materials of the earth's surface behave in accordance with fundamental principles relating to energy and matter that have been thoroughly established by the work of physicists, chemists, and others over several centuries. Interpretation of geomorphic processes requires that we apply such principles with precision and a proper degree of understanding of their meaning. Failure to do so results in theories or dogma which are at best meaningless and at worst a source of danger to man's well being.

— *Williams* (1982)

In the physical environment things exist by virtue of mutually consistent relationships, hence all the studies pertaining to environment must be based on the fact that natural components of environment are consistent in inter, and intra-relationships. Natural environments must be comprehended as a dynamic web of interrelated events, which do not emphasize on one part as being more fundamental. Thus, in analyzing environment the approach of considering it as an assemblage of basic entities possessing certain fundamental properties must be considered as primary framework.

The temperate zone has a variety of physical configuration, differing altitudinal range, diverse soil type, and dissimilar micro-climatic conditions, which together influence natural vegetation cover. Thus, considering the aims of the book, the rating criteria, which need elucidation and comprehension, are geomorphologically related items, such as soil, slope, relief and geology, and biophysical items like natural vegetation and climate, as will be elaborated in the present section. The understanding of long term behaviour in natural landforms, is ancillary upon a shift from the observation of equilibrium states to the recognition of the existence of multiple stable and unstable equilibria, the bifurcation between them and trajectories connecting them. This will rely upon a dynamic energetics approach. Not only does such an approach based upon environmental energetics offer considerable potential for investigation of the past, but it can also be applied to estimations of prospective anomalies and therefore to management.

The natural systems of the earth contain a number of interconnected subsystems, known as environmental spheres. The changes in these spheres are partly due to the natural causes, which cannot either be resisted or regulated, and partly due to direct violence of hostile anthropogenic force. Long term modifications in landform are natural eventualities associated with the natural instability of the physical environment. These processes being natural have been classified under those events where nothing can be done to influence their reduction in long term (Watson, 1983). The far greater proportion of changes are attributed to anthropogenic ignorant disregard of the natural environment. Anthropogenic intervention results in further disturbances in the existing state of equilibrium. For instance, changes in landuse, especially in highlands, results in soil deterioration and erosion, road and dam construction results in initiation of phases of persistent natural hazard, especially landslides. Thus, the relationship between anthropo-physical environment raises demurs that are not exclusively physical but all have physical aspects. An understanding of physical factors, which control natural processes, thus, is essential if the effects of natural anomaly are to be anticipated and successfully minimized.

From lithosphere to biosphere all the environmental systems respond in various ways to the flows of energy and materials through the natural systems and their interactions with anthropogenic environment. The resultant distributional patterns and movements of these natural systems, their flux and eventualities as a matter of interaction with anthropogenic environments, provide the material content of this section.

CHAPTER TWO

PHYSICAL SYSTEMS

2.1 Physical Structure

Himalaya are young mountain systems dynamic in the process of orogeny. Recent years have witnessed multitude of voluminous publications on various aspects of Himalaya by Valdiya (1973), Pande (1975), Le Fort (1975) and many others, who attempted to describe each structural zone individually, hence leading to a paltry status of information on the regional geology and conformation. A better understanding and interpretation of overall environmental peculiarities is best attempted against a background of vivid features of the area's geomorphological evolution (Krishnan, 1974). This section will demonstrate a synoptic sketch of the salient features of the generic configuration and tectonic structure of the zone.

The inquiry perimeter is comprised partially of the Lesser Himalaya, constituted of platform type depositions, supposed to be the continuation of shield components, overthrust by southward nappes and thrust sheets, and partly of the Higher Himalaya, depicting the line of great upheaval and maximum crustal attenuation and is the cardinal metamorphic belt of principally Pre-Cambrian age interceded by Granites of later ages containing Tertiary Formations (Jhingram, Thakur and Tandon, 1976) (Fig. 2.1). To facilitate the worthwhile cognition of the geological conformation of the area a districtwise appraisal of geological settings of Shimla and Kullu has been pursued, as these two districts form the basis of the empirical evidences.

2.1.1 *Geology of Chamba*

The area occupies an intermediate position between Hazara-Kashmir in the west and Spiti-Kumaun in the east. A generalized account of the stratigraphy of the region is given as follows (Rattan, 1974):

Bhalai Formation: Consists of Slates, Phyllites, Schists and Quartzite.

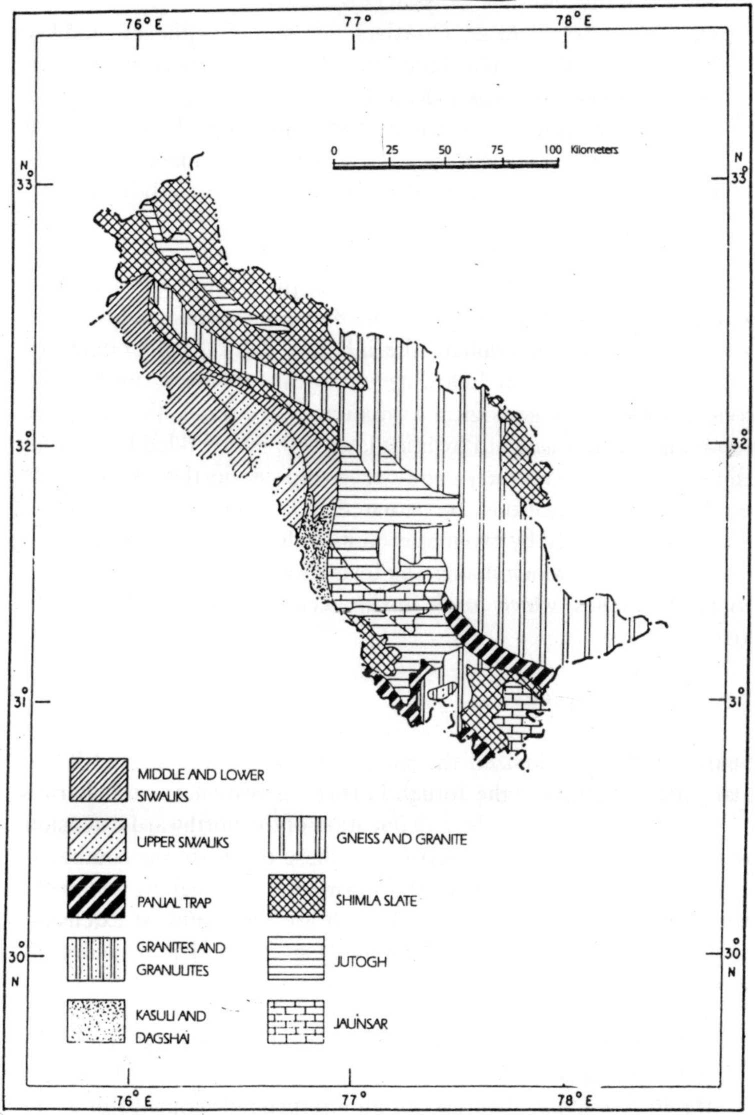

FIG. 2.1: Geology

Chamba Formation: Comprises of thick unfossiliferous metamorphosed sandy and shaley Flysch (Fig. 2.2A).
Manjir Conglomerate: Consists of metamorphosed pebbly Mudstones with intercalcated thin Slates which is recognized as the Conglomeratic Flysch (Rattan, 1974).
Salooni Formation and Kalhel Limestone: The Lower Salooni rocks belong to euxinic facies and the Upper Salooni and the Kalhel Limestone consist of an Orthoquartzite Carbonate suite (Rattan, 1979), associated by Panjal Volcanics.

With the deposition of the Manjir Conglomerates the geosynclinal basin was completely filled and the sedimentation in the area culminated. To evaluate the inherent natural vulnerability of structural conformation following empirical evidences support the conception of weak geological formation. For instance the Siul river sub-basin is comprised of Phyllitic Slates with interbedded Limestone and Dolomite/Limestone, and steep dip in northwest-southeast trend. These formations are situated in a major Kalhol Syncline which has weakened by two major faults (Dhar and Manjir), along its core of fold axis, which makes the formation prone to structural discontinuity planes, which result in slip occurrence during the shift in load factor.

2.1.2 *Geology of Kullu*

Sharma (1979) analogized the metasediments visible around Kullu and parts of Kangra to the Jotugh Formation. Srivastava and Sharma (1975) considered this to be a delineation of the northward extension of the Pandoh Synform. Nevertheless, these rocks are the southwest ward continuation of the Chamba Formation. The pebbly and gritty bed and Carbonaceous Slate of the belt are the southeast extension of the Manjir and Salooni of Chamba (Fig. 2.2A). The depiction of several Formations of the area is as follows:

Kullu Formation: Consists of Quartzites, Schists, Granitic metasediments.
Manjir Formation: Consists of an interbedded sequence of gritty and pebbly Quartzote, Phyllite, Quartzite with thin streaks of Limestone. The enlarged foundation features, inherent in repetition, depict an anticlinal fold which is in center occupied by Kullu Formation.

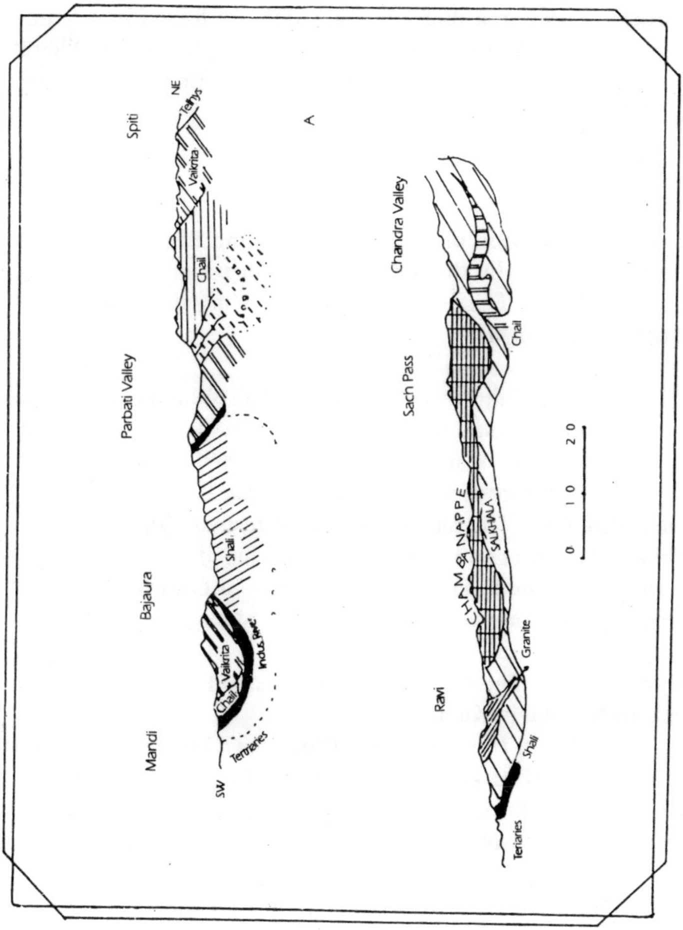

FIG. 2.2: Section Across: A—Mandi-Kullu; B—Chamba

Salooni Formation: It overlies the Manjir Formation comprising of Shales, Phyllites, Quartzite and GrayWackes. The Formation is mainly Argillaceous containing bands of Arenaceous rock units.

The rocks of the area under investigation are folded in a structurally complex Synclinorium, known as Kullu Synclinorium. There are about four overturned subsidiary anticlines with corresponding synclines on each limb of the synclinorium.

2.1.3 *Geology of Shimla*

The geological formation of this area falls in the Lesser Himalaya which in north bound by the Main Central Thrust and in south by Main Boundary Fault. The principle geological Units distinguished in this district, are as follows:

Shali Unit: This Unit is exposed in the Shali Window of Shimla hill, comprising of Quartzite, Volcanic, Calcareous and associated Shaley Members (Srikantia and Sharma, 1969). This Unit has been traced northwards upto Kashmir (Wadia, 193l).
Shimla Slate Unit: The Shimla Slate Unit rests on the Shali Unit in Shimla. Srikantia and Sharma (197l) included more rock units in it calling it "Shimla Group". It comprises of GrayWacke, Turbidites, Shales, Conglomerates, Quartzite, Slates, Limestone and basic rock.
Krol Nappe Unit: The rocks of Krol belt extend from Krol hill in Shimla to Nainital in Kumaun.
Crystalline Klippen Unit: In Shimla Klippe, two lithotectonic units have been identified known as Chail and Jutogh Thrust sheets. The rocks of this Formation are metamorphosed, and structurally folded into broad Synform by Shimla Synform.

The structural weakness observed in Shimla District is evident in lithological structure of Shalis, Basantpurs, and Quaternary and Recent sediments of terraces complied with highly fractured and jointed rock compositions, which has rendered the Tata pani-Dhar tract of Shimla District highly susceptible to landslides (Pirazizy, 1991). The Dolomites and Limestones of this belt due to inherent weakness tend to develop depressions if the natural load factor is disturbed. The scale and proportion of soil forming processes in any region are dictated by the rock categories and mineral composition and concentration (Butzer, 1976).

2.2 **Soils**

From the agricultural perspectives, the temperate regions are the most productive areas of the earth. Soils are one of the important factors contributing to this intrinsic productivity. The productivity of soil is directly decided by the rates of decomposition and decadence which are influenced by climatic parameters. The ground of a temperate forest or steppe displays a layer of undecomposed organic matter on the facade and a stratum of partially decomposed litter[1] underneath this. This layer proceeds by a totally decomposed humus. Production of fertile soils, thus, is maintained in an equilibrium between the rates of decay, recycling, and further addition of fresh material.[2] Utilization of the original forest and meadows under agricultural landuse, reduces the original biomes confined to this tracts in hillsides or semi arid regions.[3] This utilitarian process is escalating in progression of impetuous population growth coupled with the increasing need for food, which prompts the progressive use of fertile temperate soils.

2.2.1 *Soil Classification*

Different styles of soil appraisals are principally based on the variety of exigency and applications, such as planning and management of agriculture, forestry and interpretations for engineering works, besides a variety of other anthropogenic objectives that necessitate a land suitability evaluation for alternative uses. Moreover, soil assessment is pertinent in application of knowledge of soil properties and their response to different treatments and developmental uses. Soil data, however, enables evasion of hazards in demur areas and provides a basis for essential site investigations (Fig. 2.3).

The soils of the zone vary according to microclimatic conditions, perspective, and elevation range. These soils are alluvial to glacial in some tracts, and overall lacustrine in origin. The correspondence of soils to parent rocks is little, indicating a long geomorphic history, interrupted by alternative fluvial and glacial phases, as a responsible process in their evolution. The thickness of soil cover and its fertility

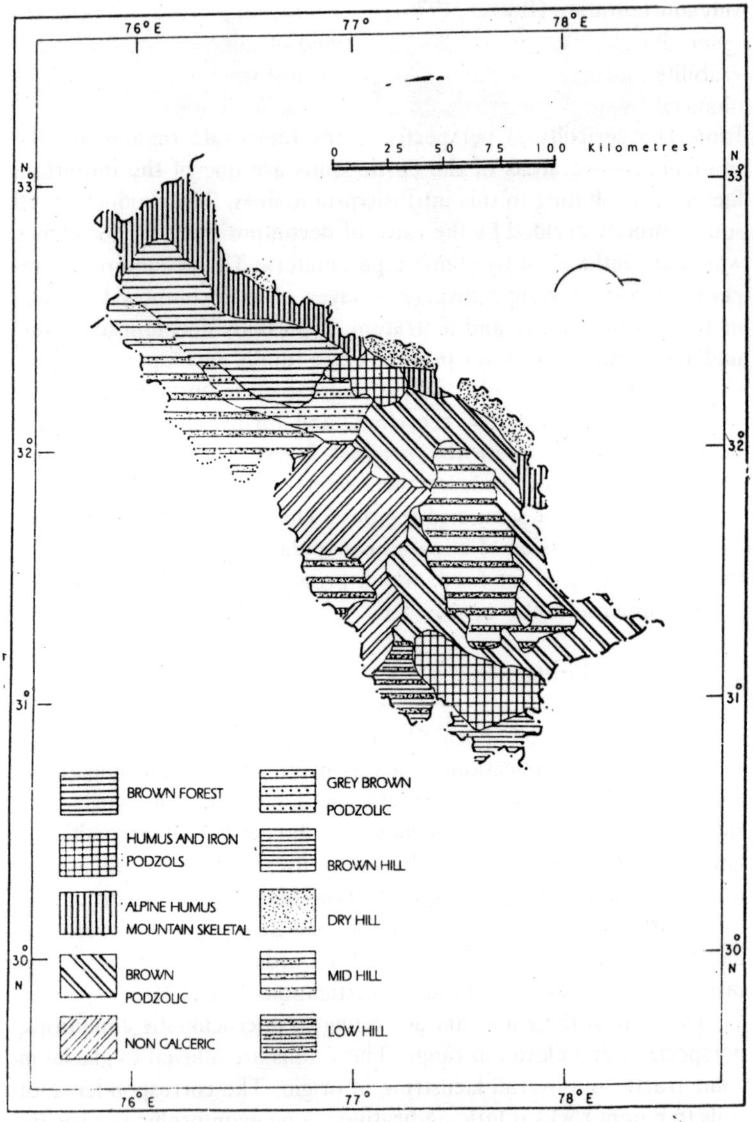
76° E
77°
78° E
0 25 50 75 100 Kilometres
33° N
32°
31°
30° N
BROWN FOREST
GREY BROWN PODZOLIC
HUMUS AND IRON PODZOLS
BROWN HILL
ALPINE HUMUS MOUNTAIN SKELETAL
DRY HILL
BROWN PODZOLIC
MID HILL
NON CALCERIC
LOW HILL

FIG. 2.3: Soils

status deteriorates with altitude. Due to steep slopes and comparatively low temperatures, which render soil forming processes slow, in higher altitudes rock beds are often devoid of soil cover. Natural vulnerability and susceptibility of soils to transformation is at present enhanced by anthropogenic interventions which perhaps, is conjecturing the supposed vicious deterioration syndrome, beyond the environmental resilience. Nonetheless, soils of the zone are classified with regards to various aspects of innate dynamism, i.e. aspect and physiography.

2.2.1.1 *Physiochemical Classification of Soils*

On the basis of Physiochemical properties the soils of the zone are broadly divided into six groups (Statistical Outline, 1988).

Brown Earths: These soils are found in parts of Mandi District. The organic matter content is moderately high with mild acidic reaction. Texture varies from loamy to sand to clayloam, with low phosphorous and medium potassium content. These soils are equivalent to Eutrochrepts, and Hapludalfs according to the new soil taxonomy.

Brown Forest Soils: These soils prevail in parts of Chamba District. They have a dark A horizon formed due to the forest vegetation. The B horizon shows alluvial clay. The soils have sandy loam to clay loam texture, with moderately high phosphorus and organic matter content. The soils are slightly acidic in nature. In order of their occurrence soils are Haplodolls and Eutrochrepts.

Grey Wooded or Podzolic Soils: This common development is spread over parts of Shimla and Kullu Districts and Karsog Tehsil of Mandi District. On stabler physiography all the soil profiles are present. Soils are strongly acidic, having a sandy loam to clay loam texture, high Potassium, low Phosphorus, and high organic matter content. Soils are Hapludolf, and Sombric Hapludolf.

Grey Brown Podzolic Soils: These soils occur in parts of Kangra District, and Joginder Nagar Tehsil of Mandi District. B horizon is well developed, marked by alluvial clay and ped surface. Heavy textures of clay loam to silty loam and silty clay are characteristics of these soils. The soils are distinctly acidic with high organic matter, medium Potassium, and deficient Phosphorous. The soils are Palendalfs, Hapludalfs, and Haplurthods.

Planosolic Soils: These soils occur in Bahl Valley of Mandi and Kullu Valley of Kullu District. The drainage is imperfect, texture fine sandy loam to clay loam and sandy clay loam, high organic matter proportion, and Phosphorus and Potassium content medium. The soils are Ocheraqualfs, Hapludalfs and Haplaquepts.

Humus and Iron Podzols: These soils are mainly confined to parts of Shimla, Dalhousie and Mandi regions. A horizon is dark coloured, and abundant free iron and aluminium concentration in B horizon. Soils are acidic in nature with a high organic matter, Potassium, and low Phosphorus content. Texture is sandy loam to sandy clay loam and clay. Soils are Haplorthods, Aguidolls, and Hapludolls.

2.2.2 *Soils of Beas Catchment*

Generally the depth of these soils vary from shallow to moderately deep. They have a pale yellowish brown to dark brown colour, and sandy to silty clay loam and clay texture. Both calcareous and non-calcareous formations are prevalent. The soils are classified as Fsamments, Fluvents, Ochrapts, and Udalfs, of which moderately fine texture, shallow to moderately deep noncalcareous soils are most erosive. Soils of basin and valleys with medium texture and thick hydromoranic surface characteristics contribute comparatively less to the silt yield of the catchment. These soils are rich in organic matter and have well developed crumb structure that renders them less erosive both because of their characteristics in parts and partially by the density of vegetation.

2.2.3 *Shimla Area of Sutlej Basin's Soils*

These soils have differing depths varying shallow to moderately deep in thickness. Their colour composition is dark brown and very dark gray. They have a loamy sand to clay loam texture, mixed with calcareous and non calcareous components in contrasting proportions. Calcareous soils are only confined to Limestone areas. Soils of these sections are broadly classified as Fsamments, Fluvents, Orthents, Ochrepts, Udalfs and Udolls. Soils of river terraces, valleys, and peidmont plains, falling in this category, are less erosive and contribute comparatively lesser amount of silt yield to the river bodies. Soils under thick to medium forest cover are dark brown to very dark gray in colour and have a thick layer of litter and humus on the surface. These soils have developed granular and crumb structure in surface and subsoils. Infiltration and percolation rates are rapid, rendering these soils less erosive to some extent.

2.3 Topography

Relief and rocks types of the zone, parts of complex and fragile abiotic environment, give it a varied appearance (Fig. 2.4). Detailed investigation of the microphysiographic features, relief, and the local hill sides, depict a promiscuousness in the distribution of relief features, that facilitates understanding of responsible processes[4], as stated by Butzer (1976):

> "Although description and categorization are . . . prerequisites of proper analysis and interpretation . . . [of] the patterns of earth surface form [they] can and should be . . . interpreted . . . in the same way as a set of phenomena in physics or biology, [and] not as . . . unrelated facts, but as a . . . system of cause and effect."

2.3.1 *General Terrain Characteristics*

A micro-level scrutiny of topographic undulation in the zone portrays a combination of highlands, lowlands, and the slopes in between. Abrupt physiographic anomaly in the zone is caused due to appearance of Lesser Himalaya in central portion, and Greater Himalaya in the north. The complexity of the terrain is further enhanced by convergence and divergence of connecting ranges, which are either terminated or get dissected by drainage network.[5] A variation in elevation from 950 meters, adjacent to southern boundary, to 4500 meters, in north, has been the main factor hindering the formation of flat plains in large number, thus, endowing the zone with numerous deep and narrow valleys. Increase in altitude has a south to north and west to east trend, in general, creating heterogeneity in physiocultural landscape of the valleys subsequent to anthropogenic settlements. Relatively simpler geomorphic features, i.e. terraces, valley, interfluvial plains, and alluvial fans, are preponderant features of the 1000–2000m. altitudinal range, while the complex features, erosion landforms with disperse population, are confined to 2000–4500m. altitudinal range and above (Joshi, 1984).[6]

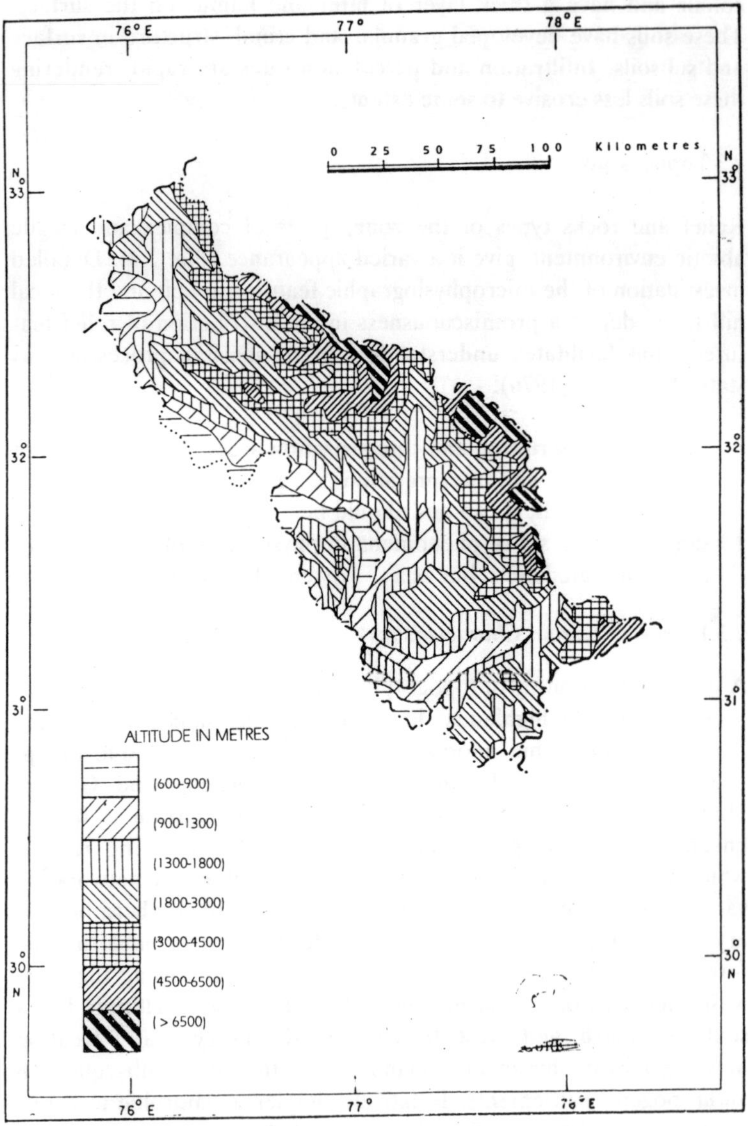

FIG. 2.4: Relief Structure

2.3.2 *Relief Type*

Elevations and inclinations are two important factors that differentiate between multifarious relief categories. On the essence of these factors two broad relief types are recognized in the zone, i.e. Mountains, and Valleys in various levels of outer and inner zones.

2.3.2.1 *Mountains*

Districts of Shimla, Kullu, and most of Chamba are situated as transitional landforms between Lesser and Great Himalaya.[7] The exact mountainous character of the zone is marked by a convexity of form (Joshi, 1984), resulting from annexed terrain of lower steep and higher gentle slopes.[8] Moreover, the presently active processes of orogeny in various stages have evolved the mountainous topography of the area, with two important ranges which are discussed below.

2.3.2.1.1 *Dhaula Dhar Range*: Dhaula Dhar meaning "grayish white" is the outer most portion of Lesser Himalayan range, which is bifurcated from the Great Himalaya in the east, stretching westward to pass to north of Shimla region, where its second branch near Kullu-Mandi boundary is situated having a mean elevation of 3000–4500 meters. Later on it terminates in the southwest part of the zone near Dalhousie. The range is a dominating land feature beyond Shimla Hills till the north of Kangra Valley. Undulating slopes and greater frequency of rock exposures of Dhaula Dhar range has rendered it a phsiocultural barrier, disconnecting the zone from the neighbouring tropical zone. In Shimla District there is a rapid change in elevation, causing irregularities in the ridges and river courses. The thrusts from the south have formed ridge types of Shimla, Chail, and Saubathu stations, causing an abrupt rise of land in Shimla Hills and narrowing the stream valleys. The Panchmuda Range with its highest peak of Chour 3647m., is situated south of Shimla (Chib, 1977).

2.3.2.1.2 *Pir Panjal Range*: Pir Panjal range a prominent water divide of the Chenab and Ravi catchments, after entering the zone from Kashmir runs through Chamba District. In its course to the north of Kullu District it divides the Chenab and Beas basins. It merges with Sri Kandh, a mid Himalayan, range to the east of Kullu. Pir Panjal range with a mean altitude of 4000–4500 meter,[9] separates the drainage basins of the Chenab and Beas from that of the Sutlej in the Inner most section of the zone.

2.3.2.2 *Valleys*

The valleys in the zone vary in size and are located at various elevations. As a general rule valleys are small, being of approximately 8.14 km. length and one-three km. width. Most of these valleys are found along the tributaries of main river systems.

2.4 Gradient

One of the major components of highland degradation with respect to anthropogenic utility impact is the steep slopes (Pirazizy, 1991). Highlands in broad descriptive sense comprise of highland, lowlands, and slopes in between them. These slopes are under continuous modification by geomorphic processes such as slides, avalanches, rockfalls, and creep. The steep slopes prevent the formation and retention of soils, due to inherent instability. Thus, terrace development on these slopes for agricultural landuse, through the application of indigenous methods by local population, is actually helpful in reducing slope instability. Therefore, on such steep slopes, soil formation can be attributed to anthropogenic impact. The inquiry perimeter is comprised of a wide range of gradient categories varying from gentle in low lying areas, to very steep on the high inclinations (Fig. 2.5). The slopes of the zone have been classified into following six broad groups:

Gentle Slopes (0–5%): These slopes hardly cover one percent of the zone's total geographical area and are confined to the flat plains of Palampur Tehsil in Kangra District.

Gentle-Moderate Slopes (5–10%): This slope type also can be termed as the minimum elevation areas. These slopes correspond to the densely populated areas of foothills and some tracts of Ani Tehsil in Kullu district. This slope category approximately covers about 5.20 per cent of the zone.

Moderate Slopes (10–15%): This category covers about 18.75 per cent of the total geographical area of the zone. This belt also hosts major anthropogenic ventures with regards to industry. North of Shimla District, and Manali Valley of Kullu District are placed under this category.

Moderate to Steep Slopes (15–20%): This category is confined to the entire Kullu Valley with exception of some patches near Manali, Bhuntar, and Larji in Kullu District, and Shimla,

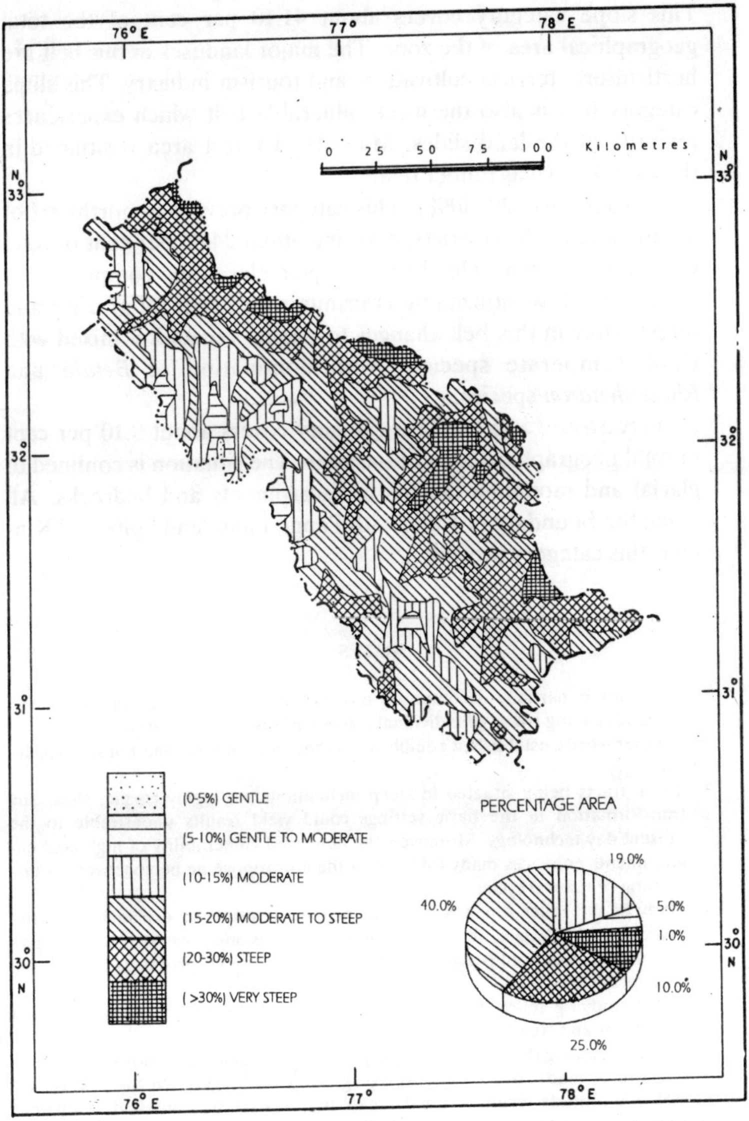
76° E
77°
78° E
0 25 50 75 100 Kilometres
N 33°
32°
31°
30° N
(0-5%) GENTLE
(5-10%) GENTLE TO MODERATE
(10-15%) MODERATE
(15-20%) MODERATE TO STEEP
(20-30%) STEEP
(>30%) VERY STEEP
PERCENTAGE AREA
19.0%
5.0%
1.0%
10.0%
25.0%
40.0%

FIG. 2.5: Slope

Narkanda, Jubbal, Rohru, Kufri, and Theog in Shimla District. This slope category covers about 41.10 per cent of the total geographical area of the zone. The major landuses in this belt are horticulture, terrace cultivation, and tourism industry. This slope category belt is also the most vulnerable belt which experiences majority of the land slides. Majority of forest area is situated in this belt, known as timber belt.

Steep Slopes (20–30%): This category prevails in northwest of Shimla and Kullu Districts, covering about 24.85 per cent of total geographical area. This belt at higher elevations comprises the richest meadows utilized by Transhumants' herds for grazing, and forest cover in this belt changes to Alpine vegetation mixed with Cool Temperate species, with domination of *Betula* and *Rhododendron* species.

Very Steep (>35%): This category covers about 9.10 per cent of total geographical area of the zone. The situation is confined to glacial and mountain heights of escarpments and bedrocks. All along the boundary of Kullu with Pangi, Lahul and Spiti, and Kinnuar this category prevails.

NOTES

1. The litter is partially decomposed material showing a change of colour and form, yet having traces of individual pieces such as twigs or leaves.
2. In other words, established equilibrium in the cycle of litter and humus production.
3. These tracts being situated in steep inclinations are highly fragile, thus, any transformation in the basic settings could yield results undesirable to the present day technology. Moreover, the perceived vulnerability of high land environments enhances many folds, from the anthropogenic perspectives, by distraction of natural cycles.
4. Study of the relief reminds one that neither description, explanation, form, process, rock, climate, the present, nor the past is adequate by itself to depict the complex balanced perspectives on the contemporary processes considering ecological parameters.
5. While studying relief the elevation and slope nature factors depict the land form types and trends of their evolution.
6. While discussing the relief and its responsible emergence processes a reference ought to be made to climate, drainage, geology, and vegetation cover, as the increase in altitude results in the decrease in temperature, rainfall, development of drainage, and the rates of stream dissection.
7. Which four or series of parallel ranges divided by longitudinal valley "Dunes", with the only exception of Kullu valley which forms a right angle to the ranges.
8. Examples of which are the terrain along the Upper Ravi River till Seraj tract in South of Kullu, and further east till Shimla-Chail part.
9. Even the off shoots of it along with its passes are as high as 3600 and above.

CHAPTER THREE

BIOPHYSICAL SYSTEMS

3.1 Phytogeography

The congenital features which give the zone a conspicuously explicit ecological character, could in a way be analyzed and comprehended through the investigation of its phytogeography (Pirazizy, 1991). Perception of environmental entity of the zone through competent knowledge of its ecology, facilitates the decision-making with regards to developmental scheme. This aspect in return assists the acquisition of information regarding ecological prerequisites of inhabitants, for environmentally compatible resource utilization, and optimum viability of inputs.

3.1.1 *Temperate Zone Phytogeography*

Due to colossal configuration of south-east and north-west alignment of ranges, and other physiographic reasons, the environment of the north-west Himalaya can be broadly demonstrated as being arid, cold, and highly elevated steppe, with low atmospheric temperature, high snow cover, and insignificant soil temperature as the basic environmental indices (Fig. 3.1).

3.1.1.1 *Sub-Tropical Forests*

These forests occur in areas with cold winters, warm summers, moderately humid climates and equally distributed annual rainfall. The species of deciduous, which restrict evapotranspiration through leave shedding process, are predominant in this zone. Broadleaved deciduous trees like beeches and maples are dominant vegetation types having a rich coverage of shrubs and herbaceous plants on the forest floor. The vegetation types between 800–100m. altitude occur only along Beas River from Mandi to Kangra Tehsil having a very limited areal extent. From 1000–2000m. altitude on south facing

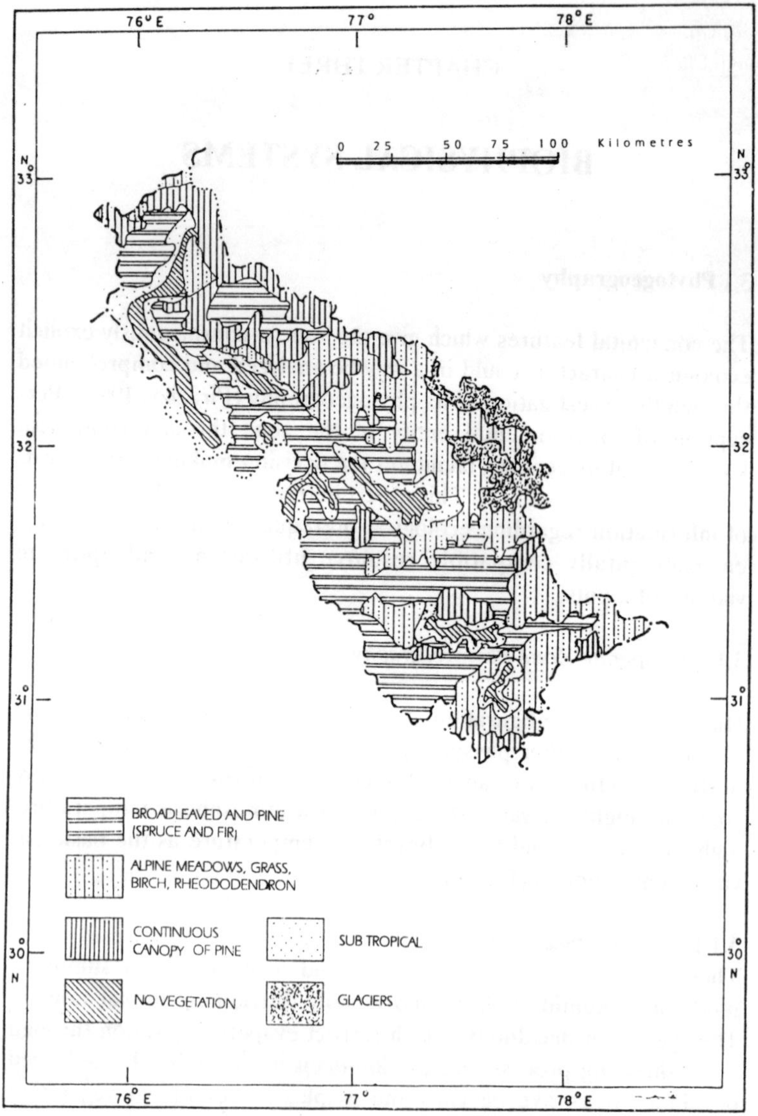

FIG. 3.1: Generalized Vegetation

slopes of Mandi, Kangra, and Shimla district, Chir pine forests dominate. These vegetation type together are known as sub-tropical vegetation types.

3.1.1.2 *Temperate Broadleaved and Pine Forests*

These vegetation types occur over conifers and broadleaved evergreen belts. These belts have poor nutrient soil type, are drought prone, and experience higher frequency of fires. Various species of pine dominates the area giving the area a boosting status of being the timber source of the zone. These forests also known as temperate vegetation types are divided into three sub-sections (Fig. 3.2).

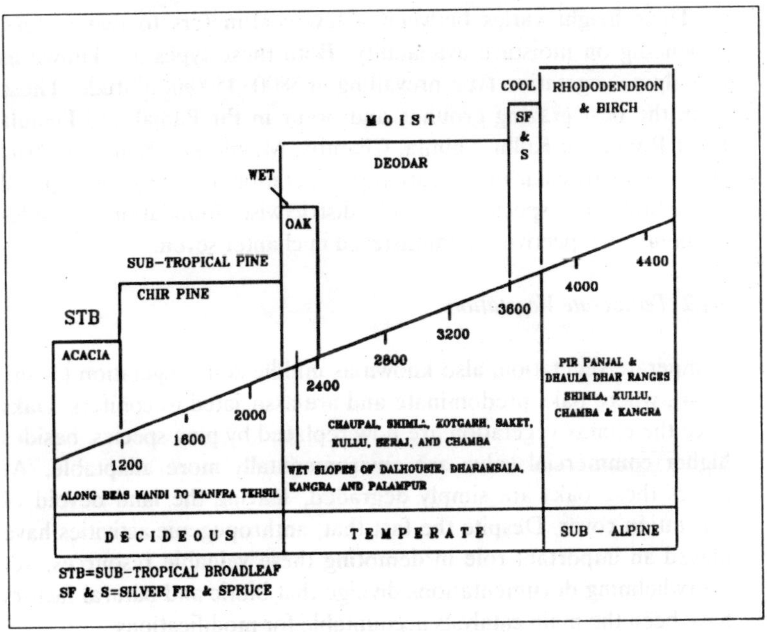

FIG. 3.2: Generalized Altitudinal Zonation Vegetation Belts

From 1500–3000 m. altitude deodar forests prevail that at many places are not pure stands. From 2500–3500m. altitude silver fir and spruce dominate which are mostly concentrated in Chaupal, Shimla, Kotgarh, Saket, Kullu, Seraj and Chamba forest divisions; while from 3000–4000m. altitude pine forests become paramount all over the zone. However, an important economic forest type under this type is oak forest, which is confined to high precipitation slopes of Dalhousie, Dharamsala, Kangra and Palampur (Pirazizy, 1991).

3.1.1.3 *Temperate Shrub and Grass Land Forests*

The shrub type of biome occurs in conditions resembling the Mediterranean dry climatic type with a meager summer precipitation. The shrubs usually have leathery leaves and range in height from one to five meters. Being aromatic in nature they are highly volatile, which in turn enhance the fire occurrence frequency. The grass type, properly known as steppe in Himalayan mountains occur in areas of insufficient moisture incapable of supporting forest growth.

Their height varies between a few centimeters to two meters, depending on moisture availability. Both these types are known as sub-alpine vegetation type prevailing in 3800–4500m. altitude. These form the best grazing grounds and occur in Pir Panjal and Dhaula Dhar Ranges in Kullu, Shimla, Chamba, Mandi, and Kangra district (Fig. 3.1). Particular investigation of vegetation, its energetics, spatial distribution on specieswise and districtwise foundation, besides economic perspective is administered in chapter seven.

3.1.2 *Temperate Vegetation*

Temperate vegetation, also known as middle zone vegetation (Joshi, 1984), where oaks predominate and are associated to conifers. Oaks once the climax vegetation are now replaced by pine species, besides higher commercial value are environmentally more adaptable. At places these oaks are simply degraded, leaving the land devoid of vegetation cover. Despite the fact that, anthropogenic activities have played an important role in demoting these valuable resources, yet overwhelming documentations divulge that biotic and natural factors have been the main catalysts accountable for modifications.

Vegetation undoubtedly is one of the vital factors in holistic appraisal of highland environments. Vegetation cover indicates the effects of physiography, soil, geomorphology, climate, and various other environmental factors (Shresta, 1977), besides anthropogenic impact (Fig. 3.2).

In belts where sufficient data regarding environment is marginal, vegetation is the best tool in bridging the void of ecological knowledge. Estimates of agglomerate productivity of vegetation and the net change in biomass is, however, achievable through analysis of temperature effect upon these environmentally multipurpose measures (Goodall, 1970). Temperature and rainfall effects are governed by climatic conditions such as seasonality and precipitation. Temperature changes have direct impact on vegetation growth, indirectly affecting soil moisture and evapotranspiration rates.[1]

3.2 **Climate**

Population of living organisms experience greater movement with regards to the patterns of climatic changes, which may cause crop failure and soil erosion. Factors deciding the climatic pattern of any zone are directly related to temperature, humidity, and precipitation. (Figs. 3.3 and 3.4).

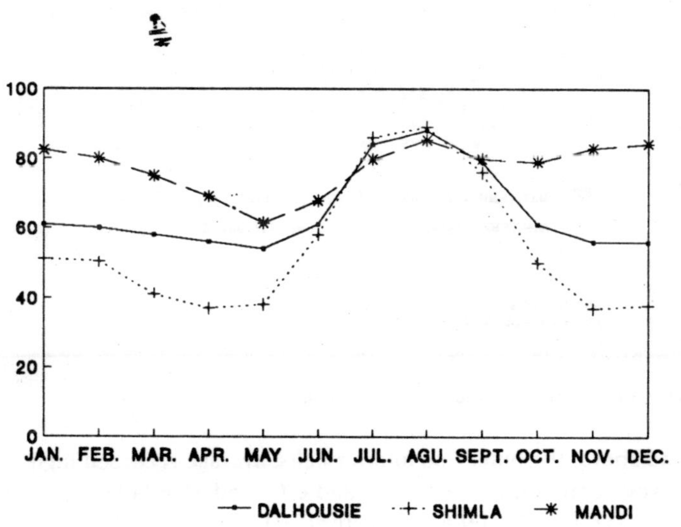

FIG. 3.3: Stationwise Relative Humidity Temperate Zone (1902–83)

3.2.1 *Temperature*

The highest maximum temperature in the zone is experienced in

June, and the lowest minimum in January. Altitudinal analysis depicts that, between 900–3500m. altitude, the average monthly temperature of summer season declines from 28°C in lower valleys of Shimla district to 16°C in inner valleys of Beas in Kullu. While at the same altitude, the winter temperatures fall from 14°C to 6°C. Almost in all the hill stations, Manali, Kullu, etc., situated at about 1500m.

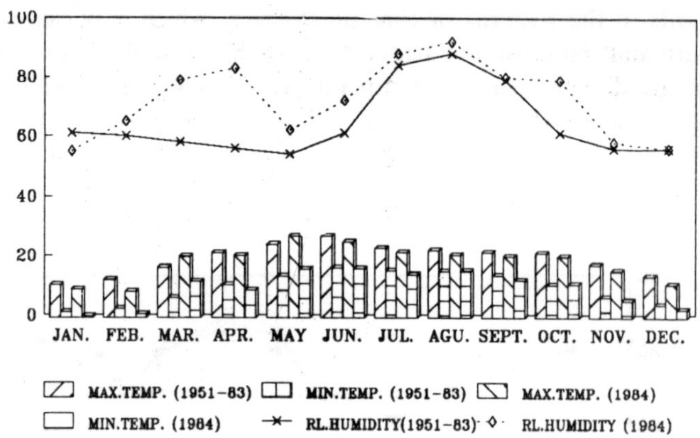

NOTE :
DATA DEPICTS THE MEAN MONTHLY FIGURES
RECORDED IN DALHOUSIE STATION.

FIG. 3.4: Temperature and Relative Humidity Dalhousie (1951–84)

and above, the outer and interior belts, average recorded highest and lowest temperatures are 22°C and 6°C, and Shimla being an exception 15.6°C and 1.9°C respectively (Figs. 3.5 and 3.6). Spatio-temporal changes in temperature distribution occurs from valley centers towards valley heads and hill tops (Joshi, 1984) (Table 3.1). The sub-tropical and temperate climatic blend could be observed all over the zone within the 900–1500m. altitudes. Above this belt cold temperate type of climate establishes between 2500–3500m. altitudes. Above 3500m. altitude sub-alpine climatic conditions are experienced. Durinal variation of temperature is greater during summer season especially on southern slopes. Length of cultivation season and growing season is decided by temperature as is the limits (absolute height), for vegetation growth. These limits are favoured with precipitation distribution, enhancing anthropogenic occupancies and activities.

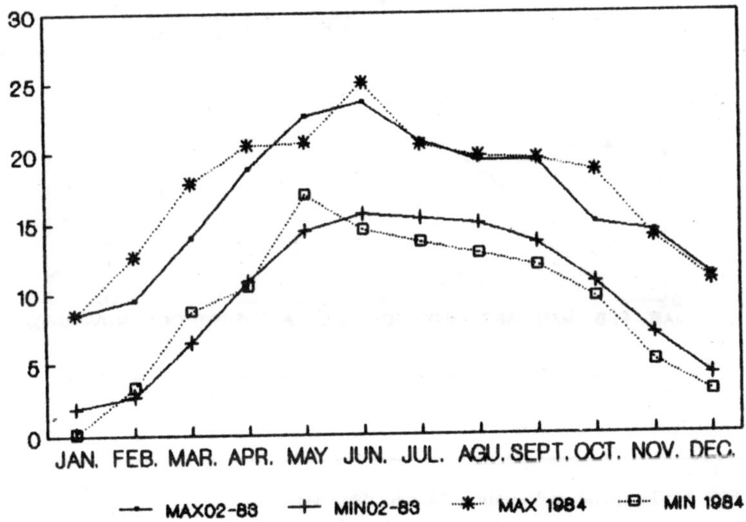

THE FIGURES FOR THE YEAR 1902 DEPICT THE MEAN OF THE PERIOD BETWEEN 1902 AND 1983

FIG. 3.5: Monthly Temperature Variations Shimla (1902–84)

3.2.2. *Humidity*

The relative humidity is broadly defined as the water retention of any specific mass of air under specific environments. The relative humidity of any area is the precipitation deciding factor of that area. The relative humidity in the zone is higher from May to September, prior and hence after it declines to the lowest. Moreover, the relative humidity of any area is decided through its active spatio-altitudinal factors (Figs. 3.7 and 3.8), For instance, proceeding from outer to inner parts, i.e. south-west to north-east, average relative humidity decreases (Table 3.2) (Figs. 3.9 and 3.10).

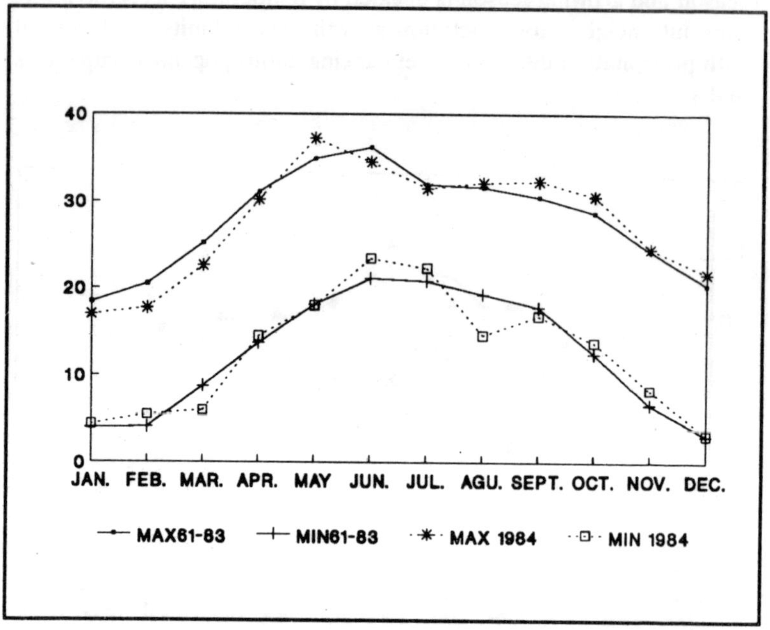

FIG. 3.6: Temperature Variations Mandi (1961–84)

During winter places near rivers and lakes, and in general all the catchment locked lands face a comparatively higher relative humidity (Fig. 3.11). Places such as Rampur and Kullu have a high winter relative humidity.

3.2.3 *Precipitation*

Within the zone the precipitation increases with increase in altitude, yet various topographic features play a determining role in its micro climatic variations. Moving north-east wards, is the belt of 500–2000 mm. annual rainfall including Kullu, Rampur, and Shimla.

However, there is a spatial decrease in precipitation distribution from outer south to inner north. Considering temperature and rainfall distribution, the climatic pattern of the zone follows a pattern of cool temperate belt at higher altitudes of Outer Himalaya including Kullu and Shimla, and cold temperate and moist temperate belt of inner lesser Himalaya including Rampur and Kullu.

TABLE 3.1: Monthly Temperature Conditions Recorded in Shimla, Mandi, and Dalhousie Stations (1951–84)

Months	Temperature			
	Maximum*	Minimum*	Maximum**	Minimum**
Shimla				
JAN.	8.5	1.9	8.4	0.1
FEB.	9.5	2.7	12.6	3.3
MAR.	13.9	6.5	17.8	8.7
APR.	18.8	10.8	20.5	10.5
MAY	22.6	14.4	20.7	17.0
JUN.	23.6	15.6	25.0	14.5
JUL.	20.8	15.3	20.5	13.6
AGU.	19.4	14.9	19.7	12.8
SEPT.	19.4	13.5	19.5	11.9
OCT.	14.9	10.7	18.6	9.6
NOV.	14.3	7.0	13.9	5.0
DEC.	11.2	4.1	10.8	2.9
Mandi				
JAN.	18.4	4.0	17.0	4.4
FEB.	20.5	4.1	17.7	5.5
MAR.	25.2	8.8	22.6	6.0
APR.	31.1	13.7	30.2	14.5
MAY	34.9	18.1	37.2	18.0
JUN.	36.2	21.1	34.5	23.4
JUL.	31.9	20.8	31.5	22.3
AGU.	31.6	19.3	32.1	14.6
SEPT.	30.5	17.8	32.3	16.8
OCT.	28.7	12.5	30.6	13.8
NOV.	24.3	6.7	24.6	8.4
DEC.	20.4	3.1	21.7	3.2
Dalhousie				
JAN.	11.1	1.9	9.4	0.5
FEB.	12.8	3.0	8.8	1.3
MAR.	16.9	6.7	20.6	12.2
APR.	21.9	11.0	21.0	9.2
MAY	24.7	14.1	27.6	16.4
JUN.	27.6	16.8	25.6	16.5
JUL.	23.4	15.9	22.0	14.6
AGU.	22.7	15.7	21.2	15.8
SEPT.	22.1	14.3	20.7	12.6
OCT.	21.8	11.1	20.6	11.3
NOV.	17.8	7.0	15.8	5.9
DEC.	14.1	4.4	11.3	2.8

* = Indicates Years (1951–83).

** = Indicate Year (1984).

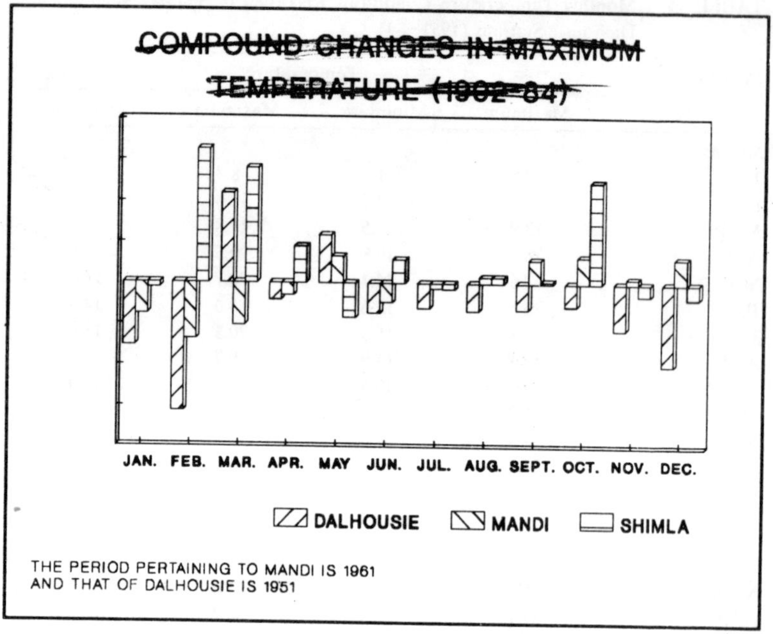

FIG. 3.7: Compound Changes in Maximum Temperature (1902–84)

TABLE 3.2: Monthly Variation in Relative Humidity (1902–84)

Months	Dalhousie		Shimla		Mandi	
	(1951-83)	(1984)	(1902-84)	(1985)	(1961-83)	(1984)
JAN.	61	55	51	48	82	86
FEB.	60	65	50	25	80	76
MAR.	58	79	41	30	75	68
APR.	56	83	37	48	69	72
MAY	54	62	38	88	61	67
JUN.	61	72	58	90	68	66
JUL.	84	88	86	86	80	69
AGU.	88	92	89	86	85	68
SEPT.	79	80	76	68	80	78
OCT.	61	79	50	41	79	68
NOV.	56	58	37	40	83	73
DEC.	56	56	38	36	84	74

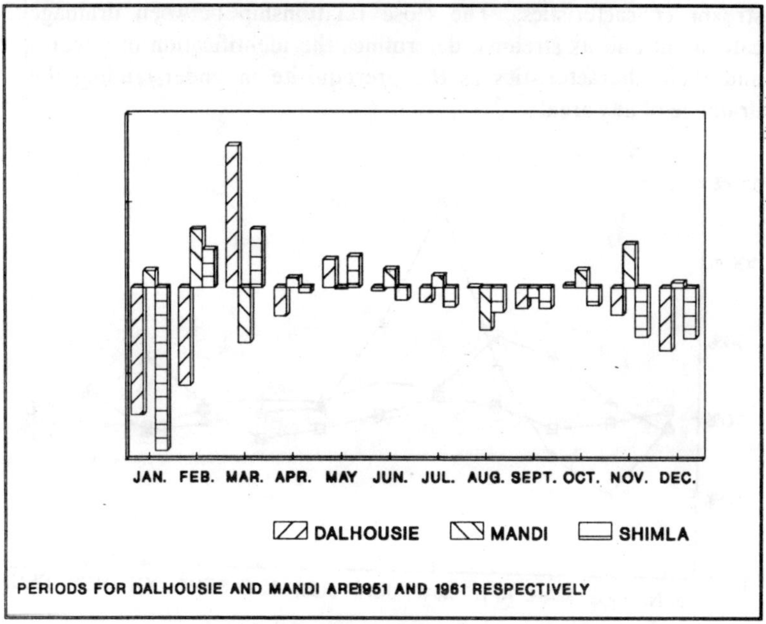

FIG. 3.8: Compound Changes in Minimum Temperature (1902–84)

3.3 Drainage or Hydrology

Any zone with a water surplus (Garner, 1974), has a downstream movement, known as draining. Streams, components of drainage system.[2] The interaction between draining water and physical features results in formation of landforms. However, various active variables complicate underactive assessment, as Chorly (1969), stated:

> ... [erosion] and transfer of sediment both occur through the action of water moving through the terrestrial phases of the hydrological cycle, and as a result the drainage catchment has also been considered a fundamental geomorphological unit".

A landscape is constituted of a mosaic of draining basins each divided by a watershed and draining to a separate stream system (Briggs and Smithsons, 1985). As drainage basins are catchment areas for the stream water, thus, any variation in them affects the stream characteristics. The close relationship between drainage catchment and its streams, determines the identification of streams and their characteristics as the prerequisite in understanding the drainage of any area.

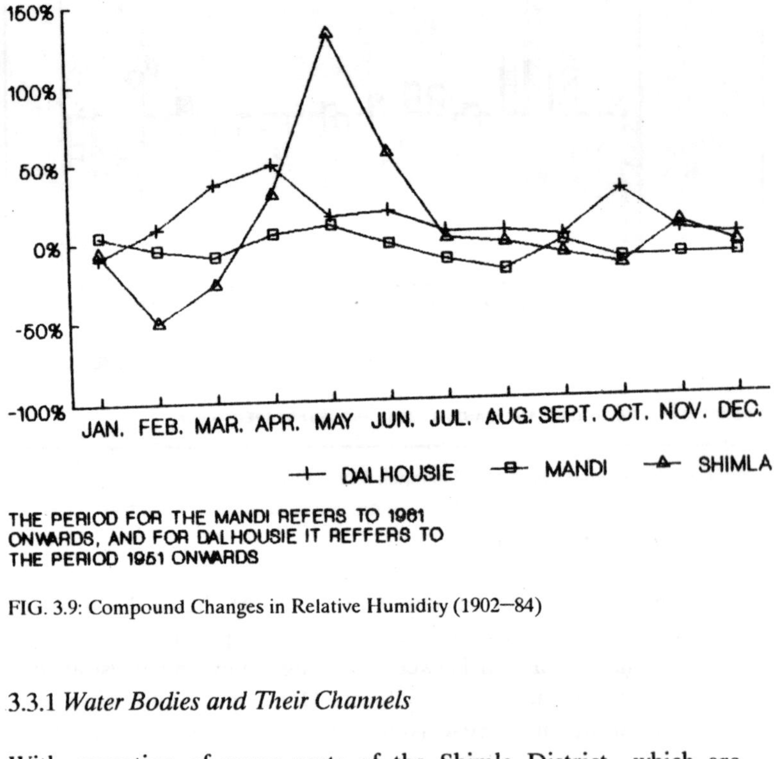

FIG. 3.9: Compound Changes in Relative Humidity (1902–84)

3.3.1 *Water Bodies and Their Channels*

With exception of some parts of the Shimla District, which are drained by rivers Sutlej and Giri, the rest of the zone is drained by two major river systems and their tributaries namely Ravi and Beas rivers. These two river systems with their tributary networks have carved two distinct catchments which are fed by precipitation. Both the river systems provide water to the Indus catchment.

3.3.1.1 *The Ravi River*

The Ravi, is the smallest of two main river systems in the zone, originating in Bara Bangahal, where glacier fed Bhadalnad taut-Gari streams joins it. The Nag Tibba range, with a south westerly orientation, joins the Dhaula Dhar and Pir Panjal ranges to form the Bangahal catchment, about 97 km. circumference (Burrard and Hyden, 1980). The Ravi, with 473 km. drainage length and 3136 km^2. catchment area in the zone, drains the southern slopes of the Pir Panjal and the northern slopes of the Dhaula Dhar range. The River has about seven major northern tributaries, some with highly steep gradients (Singh, 1971). For instance, the Bhadal rising on the north at 4877m. altitude has a fall of 60m/km. for 56 km. and the Nai rising at 5791m. has a fall of 70m/km. for a length of 48 km. The Ravi at the lowest point of Bangahal catchment has a bed height of about 1524m., and flows westwards, leaving Bangahal through a deep gorge. The Ravi flows through the Chamba district in a north west direction almost parallel to the Dhaula Dhar range (Fig. 3.12 A), till Dalhousie, where it takes a south west direction cutting a defile through the DhaulaDhar range. The average fall is 185 m/km.

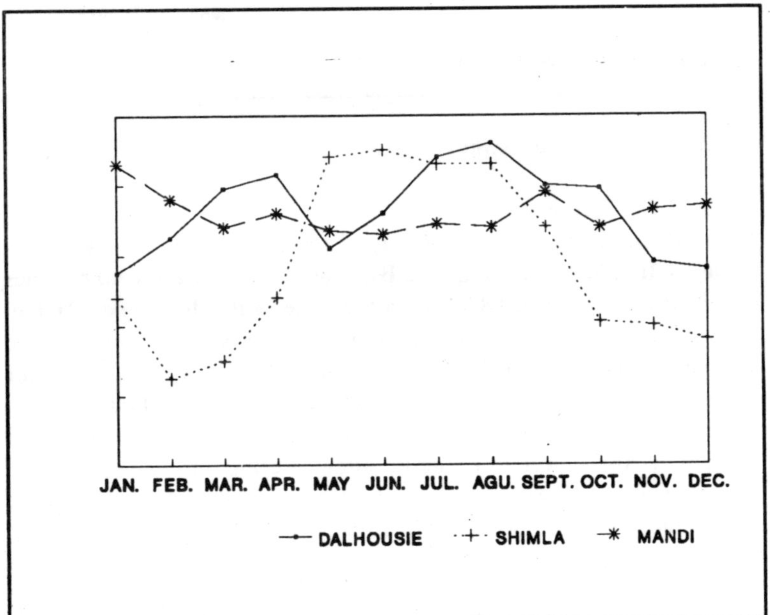

FIG. 3.10: Station-wise Relative Humidity (1984)

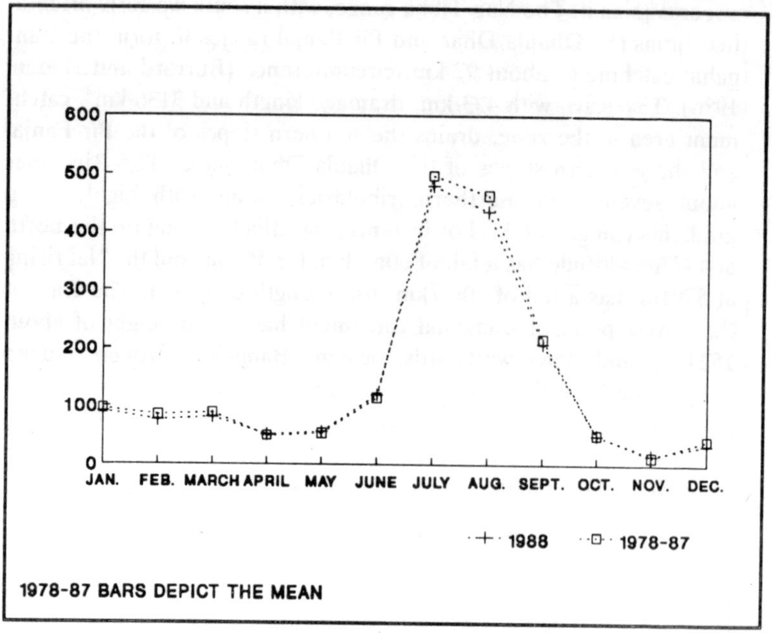

FIG. 3.11: Rainfal Characteristics (1978–88)

3.3.1.2 *The Beas River*

The Beas, rises at a close affinity of the Ravi River, near Rohtang Pass in the Pir Panjal (Fig. 3.12 B), at an altitude of about 4000m. has a drainage length of 587 km. in the zone. After following 121 km. from source along with its tributaries, the River cuts through the Dhaula Dhar range at Larji, with an average fall of 24m/km. After Larji, there is a decrease in gradient which almost is about 3m/km. Kullu district embraces the Upper Beas Basin. After Larji the River course is through a steep defile after entering into Mandi district, from where it debauchers into the Kangra Valley. Beas has about eight northern and a southern (Soan), tributaries in its 5450 km^2 catchment area in the zone.

3.3.2 *Direction of Flow*

Both, Ravi and Beas rivers in troughs the zone have a north westerly direction, which is caused due to the acute angle of the bifurcating ranges with the Middle Himalayan system. Another valid view holds that the increased height of basins in the recent geological past, and the uplift of the Middle Himalaya has been the main influencing factor in the present direction of flow in both the rivers.

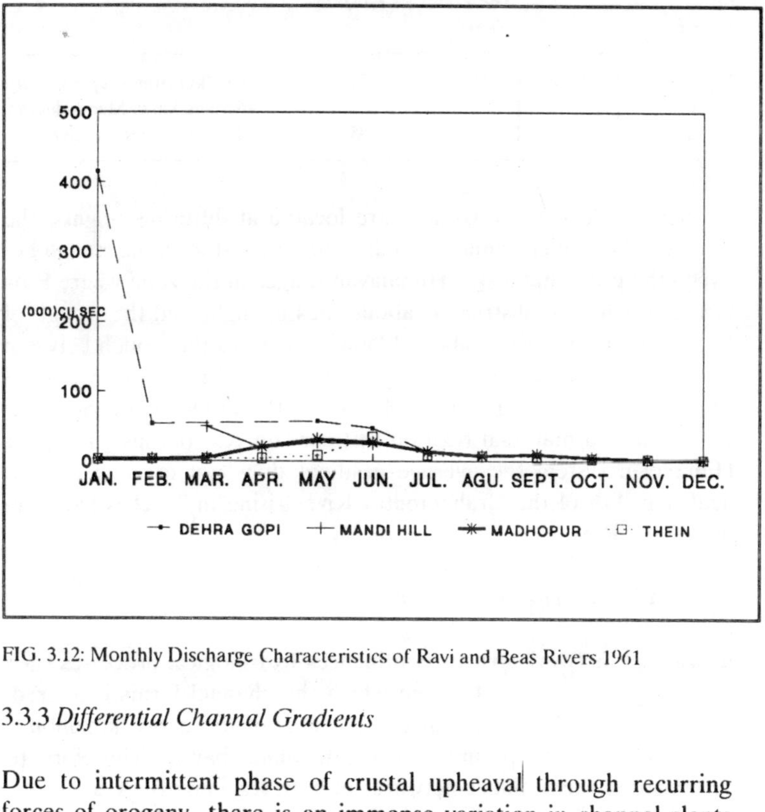

FIG. 3.12: Monthly Discharge Characteristics of Ravi and Beas Rivers 1961

3.3.3 *Differential Channal Gradients*

Due to intermittent phase of crustal upheaval through recurring forces of orogeny, there is an immense variation in channal slants throughout their profile, thus, averaging the fall estimates of the total length is a gross generalization. The normal decrement in gradient from head to base, is not a natural order in the zone, as the River systems have to cut through several parallel ranges and are confined to the troughs in the interval passing through ranges. These troughs decrease the velocity of rivers subsequent to decrease in gradient, hence, some depositional activities along their flow channals. During their flow through these troughs rivers have an average fall of 3 m/km. The Rivers cut through most of the ranges, descending the next trough with a fall of about 19–38 m/km. (Table 3.3).

TABLE 3.3: Average Gradient Estimates of the Basins

Position of Source	River	Length in The Zone (km)	Average Fall/km	Variation of Fall
In the Lesser Himalaya	Beas	587	77	24 m/km from source to Larji and 3 m/km in Mandi district
	Ravi	473	185	38m/km near the source.

Nevertheless, these troughs are located at differing heights, the exact level of which cannot be indicated. For instance, the trough between the great and lesser Himalayan ranges in the zone where Ravi flows, in Chamba district, is about 1524m. high, and the trough of Beas in Kullu district, is about 1250m., where as the trough between Lesser Himalaya and Siwaliks for the Beas River in southern parts of Mandi district is about 850m. Moreover, the gradients of the River Basins have a marginal relation to their sources, because in Upper Himalayan system the average gradient does not escalate a great deal, e.g. fall of the Brahamputra River rising in Tibet is less than that of the Ravi.

3.3.4 *Drainage Network Structure*

A drainage network structure influences hydrological processes such as sediment transport, through which the channel forms is altered. Peaks of river channel at various portions of the network exhibit a close relation to the perimeter along the main channel. Therefore, to demonstrate the drainage networks of both Ravi and Beas, a discription of their tributaries, length, and density is essencial.

3.3.4.1 *Extent and Arrangement of Tributaries*

The Ravi is fed by about 61 stream segments, differing in length and order (Table 3.4).

Moreover, major tributaries feeding the Ravi are its seven northern tributaries and their affiliated bifurcations. In general Beas is fed by main tributaries that join it from north (seven), and south (one). Overall 33 stream segments with diverse order and length comprise the Beas drainage network (Table 3.5). The last feeder of the Beas before its deboucher from the zone, is Soni, which is a second order stream joining Beas from south a little below the Pong dam site.

TABLE 3.4: Drainage Structure of Ravi

Stream Order	No. of Stream Segments	Length of Segments & % of Cummulative Length		Mean Length of Segments (km)
		Length (km)	(%)	
I	40	238.02	50.32	5.95
II	14	111.24	23.52	7.94
III	4	52.66	11.13	13.17
IV	2	51.02	10.79	25.51
V	1	20.06	4.24	20.06

TABLE 3.5: Drainage Structure of Beas

Stream Order	No. of Stream Segments	Length of Segments & % of Cummulative Length		Mean Length of Segments (km)
		Length (km)	(%)	
I	21	188.44	32.1	8.97
II	9	229.72	39.1	25.52
III	2	158.52	27	79.26
IV	1	10.32	1.8	10.32

3.3.4.2 *Drainage Density*

The comprehension of drainage catchment development is greatly facilitated through the study of the drainage density, which according to Horton (1932), is "the average length of streams within the catchment per unit area" and is expressed as: $Dd = dL/A$

where, *Dd* is drainage density, *dL* the total length of streams, and *A* is the area. The drainage density is determined by rainfall,[3] topography, infilteration capacity of terrain, and initial resistivity of the terrain to erosion (Horton, 1945).

3.3.5 *Drainage Network Composition*

The geomorphological pattern evolved through the drainage network is generally termed as the drainage composition. Drainage composition in a micro level analysis is the investigation of order, extent, and number of various tributaries draining the catchment Table 3.6.

TABLE 3.6: Drainage Composition of Ravi and Beas Basins

No. of Stream Segments	Drainage	Area km^2	Length km	Sub-Basins	Orderwise Drainage Length km	Sub-Basin Drainage Area km^2	Sub-Basin Drainage Length km
				Manali	I 10.34		
					II 5.66	273	16
				Larji	I 5.24		
					II 50.6	1274	75
					III 19.16		
I 21	Beas	5450	587	Mandi	I 30.96		
II 9					II 33.74	951	88
III 2					III 23.30		
IV 1				Dera	I 33.54		
				Gopi	II 77.9	13353	194
				Pur	III 82.56		
				Pong	I 108.36		
					II 61.82	1916	214
					III 33.50		
					IV 10.32		
				Buntuh	I 54.3		
					II 20.64	302	93
					III 18.06		
I 40				Mehla	I 51.02	1188	129
II 14					II 40.8		
III 4	Ravi	3136	473		III 21.7		
IV 2					IV 15.48		
V 1				Sungani	I 100.66	1270	160
					II 38.6		
					III 12.9		
					IV 7.77		
				Thein	I 32.04	376	91
					II 11.10		
					III 0		
					IV 27.80		
					V 20.06		

The number of stream segments draining the Ravi catchment are about 40 first order streams with a total length of 234.78 km., and 14 second order, 4 third order, 2 fourth order, and 1 fifth order stream segments, about 110.94, 51.6, 46.44, and 18.06 km. long respectively. The figures pertaining to the drainage composition of Beas catchment are 188.34 km. for twenty one first order, 227.04 km. for nine second order, 157.38 km. for two third order, and 10.32 km. for one fourth order, stream segments.

3.3.5.1 *Drainage Pattern*

Differences in physiographic and geological configuration of the zone have actuated a diverse drainage pattern. Ravi catchment has an antecedental drainage of denderic pattern till Buntuh. Then till Mehla, the pattern becomes semi parallel. Drainage pattern of Siul River, a right·bank tributary to the Ravi, is nearly rectangular till Chouhra. Beas catchment has a parallel pattern. After Dera Gopipur, Beas's course is long lateral irrigular with braided channals, resulted by constant shifting and rapid deposition of non-cohesive material.

3.3.5.2 *Discharge*

Discharge is defined as the out flow of a canal or stream. Where, the river enters another stream or body of water (Fraser, 1972),[4] or, temporal amount of water in a natural channel. Abiotic environmental factors such as depth and width of channel, alacrity, duration, and oscillation of flux and characteristics influence the discharge which subsequentle effects the physical environment (Figs. 3.12 and 3.13). Discharge, dimension, and energy of the channels tend to increase downstream in the zone.[5]

The mean discharge of the Buntuh sub-catchment (93 km. long and an area of 302 km^2.), for the year 1983 was 44221 cfs. The figure for Mehla sub catchment (49 km. long with an area of 1188 km^2., and cummulative stream length of 129 km.), was about 5185 cfs. during the same year. The Sungani sub catchment (the Suil River catchment, 69 km. long, an area of 1270 km^2., and a drainage length of 160 km.), recorded a discharge of about 8421 cfs. in 1983.

The discharge of the last sub catchment of Ravi (Thein, 88 km. long, with an area of 376 km^2., and a drainage length of 91 km.), was 10829 cfs. during the period 1983-84 (Table 3.7). The Beas catchment has five sub-basins, the first being Manali, is about 273 km^2 in area and is 16 km. long. This sub-basin recorded a mean discharge of 789 cubic feet per second in 1983. The set of hydrographs depict that, the lag time in occurence of crest increases downstream.

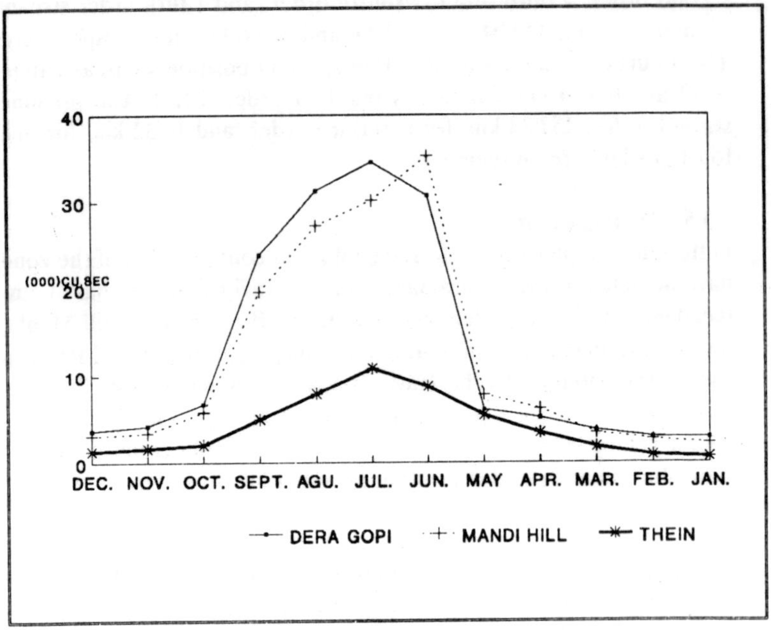

FIG. 3.13: Monthly Discharge Characteristics of Ravi and Beas Rivers 1971

TABLE 3.7: Sub Basin Drainage Characteristics of Ravi

Sub-Basin	Tributaries	Length km	Main Ravi in Sub-Basin Length km	(A) Sub-Basin Area km^2	(dL) Sub-Basin Drainage Length km	(Dd) Sub-Basin Drainage Density km	Discharge 1983 Cus.
Buntuh	Kalihen	74.94	18.06	302	93	0.31	4421
	Budhil	54.18					
Mehla	Tundan	28.38	30.96	1188	129	0.11	5185
	Halij	15.48					
Sungani	Siul	160	—	1270	160	0.13	8421
	Sal	33.58					
Thein	Siul	10.35	38.7	376	91	0.24	10829
	Seawa	7.74					

The entire period of rise and fall becomes longer downstream, leading to immense increase in discharch with the increase in catchment area (Table 3.8) (Figs. 3.14 and 3.15).

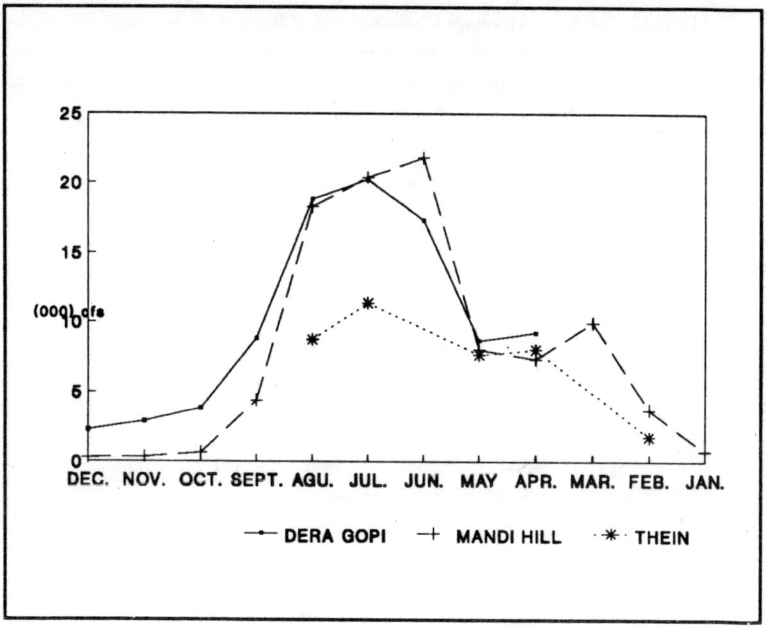

FIG. 3.14: Monthly Discharge Characteristics of Ravi and Beas Rivers 1981

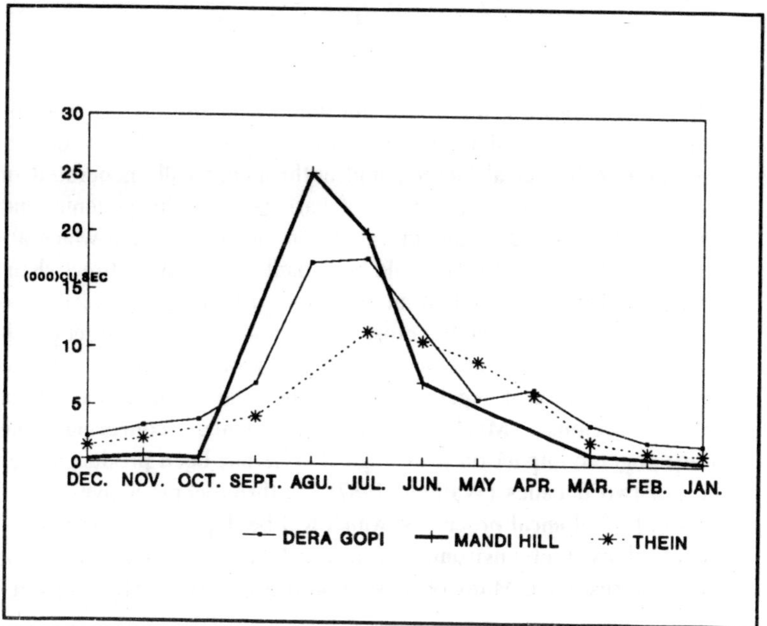

FIG. 3.15: Monthly Discharge Characteristics of Ravi and Beas Rivers 1985

TABLE 3.8: Sub Basin Drainage Characteristics of Beas

Sub Basin	Tributaries	Length km	Main Ravi in Sub Basin Length km	(A) Sub Basin Area km^2	(dL) Sub Basin Drainage Length km	(Dd) Sub Basin Drainage Density km	Discharge 1983 Cus.
Manali		–	16	273	16	0.06	789
Larji	Parbati	23.32	51.68	1274	75	0.06	7478
Mandi	Uhl	59.42	28.58	951	88	0.01	4888
Dera-	Luni	41.78					
Gopi-	Nedgal	46.44	82.56	1335	194	0.15	8438
pur	Nakher	23.22					
	Banganga	38.7					
Pong	Gaj	118.68	30.82	1617	214	0.13	9338
	Soni	25.8					

3.3.6 *Anthropogenic Impact*

While discussing the anthropogenic impact, a distinction must be made between natural erosion, operating under the influence of normal geomorphological factors, and anthropogenically generated or accelerated erosion, a corollary of variegated anthropogenic ventures. Anthropo-hydrologic interaction in the river system which according to Marwka (1974), could be broadly grouped as direct channel manipulation, watershed modification, urbanization effects, and water pollution, of which the most conspicuous has been direct physical manipulation of river channels. Modification of watershed due to alteration of natural vegetation is another significant impact upon stream regions. Much empirical evidence concering the anthropogenic impact on hydrological cycle has been produced over the last two decades (Walling, 1987). Anthropogenic activities' impact on hydrological processes, which will be depicted in the performence of existent environment, is justified in massive volume of academic research. Many of these modifications will instinctively actuate detrimental effects on anthropogenic activities, for example increase in flood hazards, damages in river regime, causin atrophy in water quality, and reduced availability of groundwater.

NOTES

1. Vegetation and rainfall tend to be closely related, so that the area of high rainfall are often characterised by a dense vegetation cover, as the vegetation enhances the infilteration and reduces run off.
2. Large number of streams and numerous variables in various environments combined with extensive range of possible adjustment to a particular event, make the prognostication of result in any single situation almost impossible.
3. Broably speaking, channal spacing within a drainage network is shown through drainage density, which in return has a close relation to precipitation, run off, and the topography of the drainage catchment (Ward, 1975).
4. Harnessing of water resources is an anthropogenic impact which, has influenced the river ecosystem, by affecting its discharge.
5. As indicated by Beckinsale (1969a), due to "the variation in [a river's] discharge", in the river regimes fro January to December, and the mean of each year from 1960—1981

SECTION III

HAZARDS AND RISKS

This section is intended to describe both the main types of hazards in anthropo-environmental highland-lowland interactive systems, their causes and consequences, and a brief theoretical basis for stability calculations and the required field and laboratory investigation. For the designing anthropogenic ventures the stabilization of hazards is acquiring immense importance, as it is unavoidable for an increasing degree to construct buildings and transportation networks in areas that were previously avoided on account of their instability. From the results the necessity to take preventive measures in hazard zones, in order to avoid further acceleration of hazards, if not their total mitigation, in relation to anthropogenic utilization of highland environments.

It is important to emphasize that this section covers a wide but not complete range of hazards, due to various limitation of research. However, the presentations contained in this section may easily be supplemented from the international literature listed in the references, which deal with hazards and consequent eventualities in most part of the fragile anthropo-environmental highland interactive systems all over the globe.

Incompleteness is apart of all the scientific field in general. Owing to the reason this section can offer opportunity to be of use to various scholars in using its findings and working towards new and even better methods for analysis and mitigation of hazards and eventual risks.

CHAPTER FOUR

ANTHROPOGENIC ENDEAVOURS

4.1 **Prelude**

It is important to understand that anthropogenic technology is capable to simulate drastic changes in physical environment in which anthropogenic environment intends to survive. This prodigious anomalies, however, have been quite intense in terms of utilization of instable highland environment for construction activities (Pirazizy, 1991). This aspect has led to instantaneous disruption of the dynamic balance of natural systems, which have been established over along period. The adjustment to these anthropogenically imposed modifications has not been adequately anticipated. One of the major modifications resulting in drastic consequences has been the utilization of highly instable highland gradient for road and dam construction with landslides of differing magnitude and frequency as aftermath.

4.2 **Instability Premium**

Landslide prone slopes tend to posses certain distinctive features, including various combinations of steep slopes, high slopes, high porewater pressure, thick and deeply weathered soil cover, undercutting of the base of the slope, and weak material outcropping below stronger material. Of increasing significance, however, is the role of anthropogenic ventures. This impact is resulted by various characteristics. These characteristics can be classified as:

- Excavation at the base of a slope, creation of steep slope, and removing the basal support.
- Diversion of excessive water into slope.
- Clearing vegetation cover, which in return increases the slope degree. These vegetative cover, would normally have resisted

shallow failures, not only through binding effects of roots, but also through the absorption and evapotranspiration.

All above factors are directly controlling factors of shear resilience and stress, through which slope stability is affected. Today road designs intend to have a smooth line which stipulates a cut through steep slopes, that are almost at periphery of equilibrium (Veder, 1979). Despite the fact that, gentler slopes are highly preferable for road construction, yet land acquisition and operation costs stress the execution of such construction feasibilities, dictating construction on steep slopes. For instance road-bed and cut-slope are, in Himalayan case often, at 25–30 degree. Seasonal saturation and weak vegetative cover, together, prone these slopes vulnerable to failure ensued by intense erosion, especially in termination of construction work. Most of the landslides along the roads in the zone occur in the altitudinal belt of 1200–2300m.

About 80 per cent of landslides have been recorded in the mentioned elevation, where weak geolithological structures are constraint on construction activities. The road net work in the zone exhibited an abrupt increase from 6000 to 6800 km. only two years period i.e. 1977–78 till 1979–80. The figures apply only to main roads, besides which many kilometres of road has been constructed. Erroneous planning and obdurate execution of construction ventures of most of these roads, has reduced stability of the hill sides through which these roads pass, generating massive debris quantity, which is thrown down the slopes. This sliding of debris causes irreversible damages to natural vegetation cover in its path of flux and amplifies the erosive power of the streams by charging their natural flow. According to estimates of Singh (1985), only one kilometre of road generates about 40–80 thousand cubic meter of debris. The new construction sites, become damage prone, through mass movements of various types, especially landslides, during the rainy seasons (Pirazizy, 1991).

The existing roads coverage in the zone, is about 6800 km. (Fig. 4.1), which must have generated not less than 400 thousand cubic meter of debris. According to Boarder Road Organization an annual rate of about 750 m^3/km. of debris is cleared from the roads under their function in the state of Himachal Pradesh (Patnaik, 1981). Construction of roads and engineering cuts of the slopes which according to Valdiya (1987), are three to five meter high and four meter wide, produce sizable volume of debris. Estimates depict that for one cubic meter of rock, about two cubic meter of debris is produced. Haigh (1984), estimated that in general excavation destroys one tree in every 500 tress on the slope. Despite the fact that the extent of damage to vegetative cover of the slopes in the zone is staggering, yet, even a gross estimate proves a large scale damage during the last few decades of rapid road construction and expansion. Considering only the average, about 500 $m^3/km/yr$ of debris is generated from the existing roads, which is not a merely conceptual estimate, the zone with about 6800 km. road length in aggregate generates about $34x10^5$ m^3/yr of debris in addition to tonnes of debris produced subsequent to construction of new roads. The increase in road length and number in the zone was a function of two major programmes, i.e. Sensitive Boarder Area Roads and Hill Region Roads. This expansion of roads transgressed on higher altitudes possessing scanty vegetation cover and weak geolithological structures, hence, corollary of immense landslides (Pirazizy, 1991).

The altitudinal dissimilarity of the origin and destination areas is considered to be an important component eliciting landslides. The main road systems in the zone originate and terminate at different elevations, with a diverse range of height of passage (Table 4.1). The roads with low origin and high destination, vice versa, are the belts where large number of landslides occur. Despite the construction incompatibility with fragile ecosystems of mountains, most of the roads have been widened to accommodate two way traffic.

Construction of roads and engineering cuts of slopes which according to Valdiya (1987), are about three to five meter high and four meter wide, produce sizable volume of debris. Estimates depict that for one cubic meter of rock about two cubic meter of debris is produced. Haigh (1984), estimated that in general excavation destroys one tree in every 500 trees on the slope. Despite the fact that the extent of damage to vegetation cover of the slopes in the zone is staggering, yet even a gross estimate proves that a large scale damage has ben inflicted on natural vegetation cover during the past few decades of rapid road construction and expansion (Pirazizy, 1991) (Fig. 4.2).

4.3 **Perimeters of Instability**

The aim of the present excerpt is to facilitate the depiction of the overall extent of road construction and consequent types of degradation hazards through provision of relevant statistics.

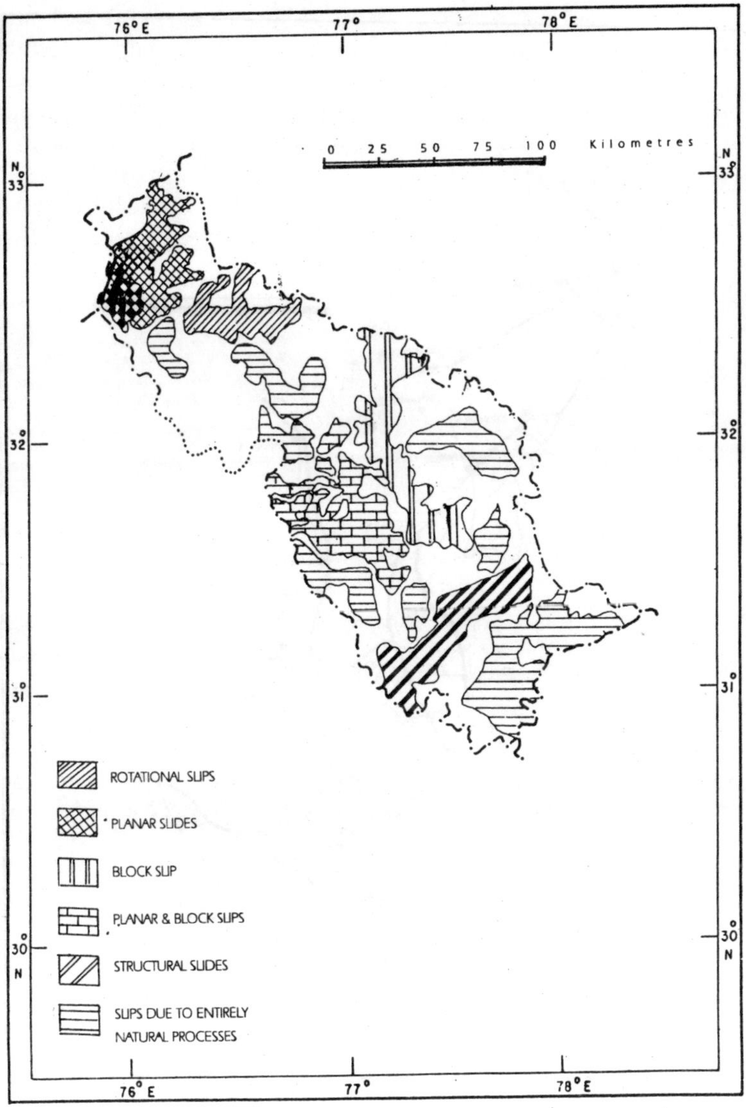

FIG. 4.1: Landslide Risk

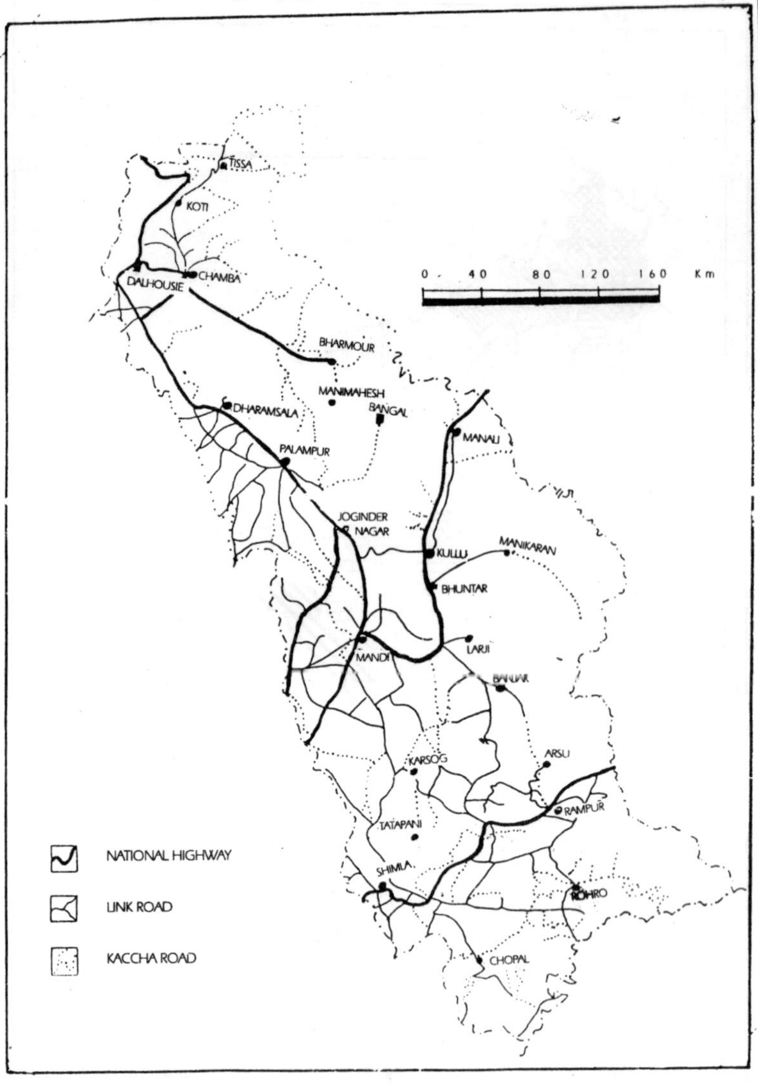

FIG. 4.2: Status of Roads

Besides, this section intends to illustrate the most representing conditions of degradation prevalent in the zone. The specific criterion used is to elaborate on the movement of slopes which depict the aim of the investigation into this pressing environmental issue. The aim is associated with an acute task of stability analysis, prevention, and avoidance of risk, and obstructing further degradation.

TABLE 4.1: Road Length

Origin	Height (M)	Destination	Height (M)	Length (km)
Dharamsala	1250	Mandi	800	47
Dharamsala	1250	Kullu	1176	214
Dharamsala	1250	Shimla	2213	332
Dharamsala	1250	Chamba	995	192
Dharamsala	1250	Kangra	560	18
Dharamsala	1250	Triund	2827	17
Dharamsala	1250	Palampur	1219	35
Palampur	1219	Kangra	560	40
Palampur	1219	Pathankot	500	115
Palampur	1219	Kullu	1176	195
Palampur	1219	Mandi	800	96
Palampur	1219	Shimla	2213	259
Shimla	2213	Samkat Mochan	1873	7
Shimla	2213	Mashobra	2149	13
Shimla	2213	Kufri	2633	16
Shimla	2213	Fagu	2510	22
Shimla	2213	Tatapani	655	51
Shimla	2213	Chail	2250	90
Shimla	2213	Kasauli	1846	77
Shimla	2213	Kullu	1176	240
Shimla	2213	Manali	1829	280
Shimla	2213	Nahan	2588	140
Shimla	2213	Kalpa	2705	260
Shimla	2213	Chamba	995	502
Shimla	2213	Dalhousie	2093	436
Shimla	2213	Mandi	800	156
Shimla	2213	Teradevi	1851	11
Shimla	2213	Rohru	1149	115
Shimla	2213	Narkanda	2709	94
Shimla	2213	Kotgarh	1830	82
Shimla	2213	Naldehra	3044	22
Chamba	996	Pangi	5425	137
Chamba	996	Dalhousie	2093	50

(Contd.)

TABLE 4.1 *(Contd.)*

Origin	Height (M)	Destination	Height (M)	Length (km)
Chamba	996	Mandi	800	334
Chamba	996	Manali	1829	470
Chamba	996	Khajiar	2141	27
Dalhousie	2093	Kalatope	2440	9
Dalhousie	2093	Bakrota	2085	5
Manali	1829	Rohtang	3979	51
Kullu	1176	Mandi	800	56
Kullu	1176	Saluni	1835	67
Kullu	1176	Brahmaur	2141	67
Mandi	800	Jogindernagar	1300	56
Mandi	800	Barot	1900	75
Mandi	800	Sundernagar	1000	26
Mandi	800	Rewalsar	1300	19

Most of the landslide occurrence, especially those that have caused extensive damage to life and property, are recorded along the highways and main roads of the zone (Table 4.2).

TABLE 4.2: Status of Roads

District	1977-78			1978-79			1979-80		
	Met.	Unmet.	Total	Met.	Unmet.	Total	Met.	Unmet.	Total
Chamba	165	691	856	176	681	857	193	719	912
Kangra	613	1244	1857	642	1358	2000	703	1370	2073
Kullu	150	435	585	184	411	595	199	407	606
Mandi	425	888	1313	453	939	1392	478	995	1473
Shimla	663	841	1504	717	896	1613	719	1014	1733
Zone	2016	4099	6115	2172	4285	6457	2292	4505	6797

Source: PWD Himachal Pradesh.

The causes of landslides, besides natural factors, are more consequential to type and pattern of anthropogenic influence on their generation. As the developmental, industrial, and construction ventures, such as settlement, industry, and tourism are encroaching into higher reaches, due to population pressure, mass movements are encountered directly, being considered as anthropogenic and becoming more serious (Pirazizy, 1991).

Three major highways of the zone which, besides being the life vain of the economic activities, are the scenes of inclement landslides, e.g. Pathankot-Chamba Highway, passing through Dharamsala, Palampur, Dalhousie, Kalka-Rampur Highway, passing through Dagshai, Kandaghat, Shogi, Shimla, Mashobra, Theog, Narkanda, and Kumharsain, and Mandi-Manali Highway, passing through Pandoh, Larji, Kullu, and Wagar.

A detailed examination of Kalka-Rampur highway reveals that from Kalka till Shogi the path is over the elevation zone of 1200–1800m. on the geological formations of Shales, Limestones, Siltstones, and Orthoquartzites of varying lithology. This section of highway has experienced about 54 zonal slides during the last two years. The concentration of landslides in this lithological zone is 1.49. The next section of the highway, Shogi till Narkanda, the path is through Siltstone, Slate, Quartzite, and Phyllite geological formations, experiencing about 63 zonal landslides with a concentration index of 4.72. The last portion of the highway, Narkanda-Rampur, passes through basics and intrusives with a stabler structure. Mandi-Manali highway is less stable than Kalka-Rampur, with low shear resistance of geolithological structures till Larji. This section passes through Limestone, Dolomite, and conglomerate structures of Larji Shalis and Manjir formations. The concentration index of the slides in this belt is 4.34 . From Larji till Manali the highway becomes stabler with the instability peculiarities of the road side hill slopes. The number of landslides increased substantially after 1967, which coincides with the tenure of the 4th, 5th, and 6th, five year plan (Mukarji, 1986). From 1951 to 1981 about 75 per cent of the landslides on Mandi-Manali, and 96 per cent of the landslides on Kalka-Rampur highways are the product of post-1967 era.

In 1960s and 1970s the haste in road construction, without proper attention to environmental stability, was to establish the network, being considered as the life line of development. Since then the major task has been the maintenance problem. These problems are socio-political (Planning Commission, 1982a), which even disturb the technological order of the construction.

4.4 **Premiers of Degradation**

Anthropogenic enterprises and occupancy of the zone play an important role in multitude rate of accelerated highland degradation. Field observations suggest that, deforestation disproportionate to afforestation, overgrazing, urban expansion, and increased transportation network are the main anthropogenic causes generating landslides, which are common features all over the zone in the form of scarred landscapes (Pirazizy, 1991).

Increase in the length of the existing roads as a major component of regional developmental theorems, besides strategic importance, has amplified the number of slides due to malevolent execution and unpremeditated planning. For instance, the method of blasting actively employed as a means of excavation, results in internal disintegration of regolith and rocks, reducing stability already weak geolithological structures, thus promoting further slides.

Most of the landslides are concentrated between 1200–2500m. altitude. About 18.35 per cent of landslides occur in the altitudinal belt of 800–1600m. The percentage of landslides in the altitudinal belt of 1200–1600m. are 41.77 per cent, in 1600–2000m. are 8.23 per cent, and in 2000–2500m. are 30.38 per cent. Lower elevations of between 900–1600m., which comprise of weak geological structure of Shales, Limestone, Dolomite, Slates, Phyllites, and crushed Quartzites, experience about 60.13 per cent of the total landslides. Strategic and developmental importance of road construction has been the main factor in amplifying the number of slides due to malevolent execution and unpremeditated planning. Fore instance, the method of blasting actively employed as a means of excavation, results in internal disintegration of regolith and rocks, reducing stability already weak geolithological structures, thus promoting further slides. Inorder to safeguard environment from unwanted impact of road construction and expansion the following general evaluations must be carefully conducted.

4.4.1 *Chamba Slides*

In Chamba district, the valley slopes are mostly rocky with talus accumulation along mantles. Prominent terrain and structural features of the district are, rocky steep slopes, moderately outward dipping beds in Slates/Phyllites, open tension relief joints along valley slopes, and open symmetrical synclinal/folded settings of cross fractured dips into the valleys. Water pressure along the joints during rains, is the major factor prompting slips and mass movements.

Major landslides occur along the Chamba-J&K, Mandi-Lahul (Rohtang Pass), Chamba-Tissa (mostly on the second half of the road with almost 30 landslides a year), Dharamsala-Chamba, and Pathankot-Dalhousie roads, which have exposed infra-shali, Paleozoic-Pre-Cambrian Chail Phyllites in thrust contact with bands of gneiss at the base of Paleozoic-Pre-Cambrian Chail Phyllites and Slates. At many places, especially milestone 42/1, the saturation of the highly crushed fault zone material leading to their consolidation under load pressure results in failure of the road by almost maximum of two meters. Dalhousie cantonment road landslide, for example, faces a planar creep in the Chail, Phyllites and Slates, and consequent debris-slides, till now have devastated many settlements, especially on the Balun spur. Mass movements in Chamba district are mostly of two major types (Gosh, 1980). Rotational slip in cohesive debris/talus material, common along the Siul River, Chamba-Tissa road and Planar slides in bedded rock slope, common along Baledh valley (Slate/Phyllite), and Kalhel valley (Limestone/Slate) (Fig. 4.3).

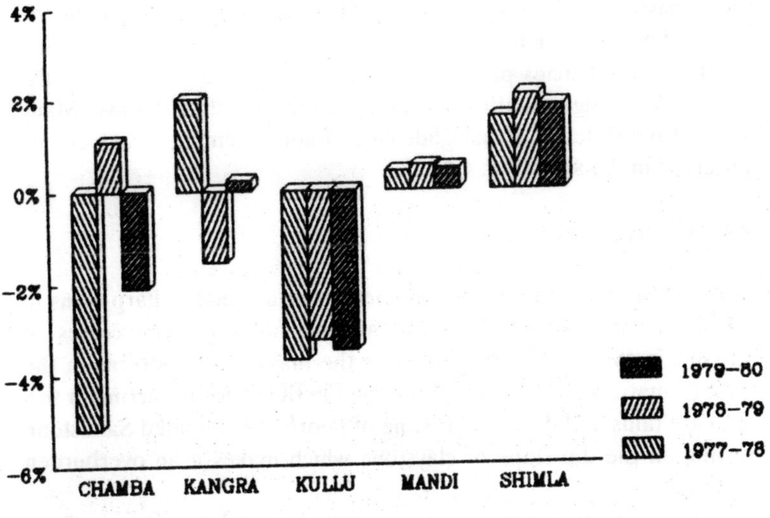

FIG. 4.3: Districtwise Variation in Total Road Length (1977-80)

4.4.2 *Kullu Slides*

Most of the slides in Kullu district occur along Kullu-Mandi road. The major landslides are recorded to happen, annually, in the vicinity of 7 mile away from Kataula, and 13 miles away from Pandoh-Nagwin on the same road. The jointed Chail Phyllites making up the structure of these highways are affected by deep, gravitational block-slip, which causes open glide-crack formations in the hill masses through which these highways pass. The base erosion of spurs, forming abut of these runs with the Beas River, are the main causes of such potentiality for deep seated block-slips and land slides (Fig. 4.3).

4.4.3 *Mandi Slides*

Every year many landslides are recorded along all the link roads in the district, which are caused due to unplanned cutting of the precipitous high degree slopes on the weak geolithological structures, with literally no retaining walls (if existing usually < 1.5m.).

There are two major areas where large scale landslides are recorded i.e. Maniue and Pong dam site. In these areas an alternating sequence of soft sand-rock and clay-shale of Palio-Pleistocene (Krishnaswamy, 1980), upper Siwalik formation is exposed in the gentle dips to the left abut on the Beas river. Marked planar and block slip tendencies prevail in these areas that, are guided by the plastic clay-gouge-mantles occurring in the clay-shale bands. Many scars of old slides of planar glide type of not so remote a past, can be observed in the area (Fig. 4.3).

4.4.4 *Shimla Slides*

Between Shimla and Bilaspur districts there are many sharp turns in alignment of ridges due to folding and faulting. The corses of streams in this area is controlled by the major joint patterns in the rock formations. The material involved in the slides occurring in this area are talus and debris consisting of poorly consolidated Sandstone resting on the dip slope of claystone which makes it an overburden material.

On the steep slopes of the area usually circular failure on fractured and weathered sandstones occur (Raju and Jalote, 1980). Shimla slides are well known over the last 25 years. The debris slides and slump-failures on the northern slopes have been affecting the construction activities. Geologically Chail Phyllites are overlain by a thick accumulation of clayey debris and the bed rock over burden interface deflecting northwards. This geological environment facilitates the down ward movement of slope forming material ingress of surface water into the overburden has disturbed rock configuration and bed rock structures, hence, failure of slopes (Fig. 4.3).

4.5 Landslides Correlates

The term landslide is defined as the collective term for the slope movements of the mass movement category excluding creep and subsidence. However, the generic term, landslide, is the sum of all down slope movements of material due to shear failure at the boundaries of the moving mass (Skempton and Hutchinson, 1969).

During the last 20 years, the zone has experienced construction of many hundred kilometers of mountain roads (Fig. 4.1). Most of these roads were constructed due to the demand of national security, which needed road construction as cheap and quick as possible. This aspect led to concentration of construction activities on complex geolithological units across highly precipitous mountain slopes. Construction of these roads made the access to remote forests, increasing illegal large scale forest removal (Haigh, 1982). Thus, large scale environmental disturbances and a persistent problem of slope failure, especially along the roads, as stated by Tejwani (1987).

> "If currently any single source of sediment generation, instability of hill slopes, [anthropogenic] cause of landslide, erosion . . . and forest degradation is to be identified, it will . . . be the vast network of roads [in highlands]. . . ."

Geological formations such as Slates, Shales, Limestones, Phyllites, and Quartzite, in the faulting belt of the temperate zone, are of crushed nature, and abound in the zone, being responsible lithological factors in landslides. Steep slopes, a function of structural dips, combined with deforestation (Mukerji, 1986), are highly conducive physiographic conditions for enhancing hydrologic shear intensity, thus, reducing stability the zone over a major part, leading to numerous large scale slides (Pirazizy, 1991).

Among the major anthropogenic correlates in the zone as stated earlier, are disproportionate urbanization, hasty and malexecuted transport network construction and expansion, and heavy industrial activities location disregarding the geolithological weakness prevalent over the highly elevated and variegated altitudes. The geological environment of the zone is such that, landslides are common features. Considering the overall terrain conditions and structural conformation of the zone, the prominent slope failure elements influencing the construction activities are, steep bed gradient of the rivers and active bed erosion during peak discharges, steep valley slopes controlled by joints, interbedded soft and hard rocks, and outwardly dipping divisional planes-bedding, Schistosity, shear connections, and joints.

The usual commencing point for identification of landslide prone areas is to record (map), the position and character of any existing landslides. This assessment proves useful in following major manners:

- To illustrate the perimeter which could experience reactivated landslide.
- To appraise sites where landslides have already taken place, in order to comprehend and analyze the causes leading to landsliding in those areas;
- To identify those areas which are close to the threshold of movement or their slope factor of safety is close to unity.

To accomplish this objective which led to construction of (Fig. 4.2), landslide inventory mapping, geomorphological mapping, potential erosion hazard and susceptibility mapping, to be elaborated in proceeding chapters, were developed and perused. Some of the techniques shall be discussed in detail in following chapters.

Features demonstrating slope movements range in size from small terracettes indicating creep, upto large slides (Cooke and Doorkamp, 1990). Using Varnes (1958) classification of masswasting, following slide types have been incorporated in the mapping procedure and appraisal techniques. Landslide occurrence, though not bound to administration boundaries, have been recorded in different districts of the zone. If observed collectively a pattern of landslide prone zonal pattern emerges, which besides being useful inventory in hazard avoidance and minimizing risk, can on minute scrutiny yield substantial information on the causes, types, magnitude, frequency, and emerging scenarios of management and development, on the landslides in the context of microlevel areas. Following classification of slides (Fig. 4.4), is a modified and simplified form of classification by Varnes (1978). The simplification was because of the peculiarity of the research perimeter.

Movement		Material			
		Rock		Soil	
		Planar	Rotational	Planar	Rotational
Slide	Small Area	Block Slump	Slump	Block Glide	Block Slump
	Large Area	Rock Slide	—	Debris Slide	Natural Slips

Source: Modified from Varnes (1978).

FIG. 4.4: Classification of Slides in the Zone

Landslide occurrences though not bound to administrative boundaries, have been recorded in different districts of the zone. If observed collectively a pattern of landslide prone zonal pattern emerges, which besides being useful inventory in hazard avoidance and risk minimization. can on minute scrutiny yield substantial information on the causes, types, magnitude, frequency, and emerging scenarios of management and development, on the landslide, in the micro-level context.

4.5.1 *Translational Slides*

These are the most common form of landslides occurring in soils. They are always shallow features and have essentially straight slide planes with some curvature towards the crown and hence some rotational movements in the zone. These slides are mostly manifested in the form of rock slides, block glides, debris slides, and mud flows. With relatively large undeformed blocks of soil sliding downhill, these slides may become an avalanche of small blocks and wet debris in midslope, and a thoroughly liquefied flow at the base of the slope, especially if descending into a river body. These slides always occur during heavy rain, when the tension cracks are filled and field capacity of the soil exceeded. In brief these slides involve a more or less inclined (planar) surface.

4.5.2 *Rotational Slides*

These have curved failure planes and involve rotational movement of the soil mass. These are common in soft rocks such as shales, mudstone, and overconsolidated clays. The occurrence of these slides is entirely natural, especially once the slope foe is undercut by river. Slumps may be single rotational failures with shearing occurring fast on a well defined curved surface. Some slumps, especially those in clays with hard impervious underlain strata, and porous overlain rocks, develop multiple rotational slides, with two or more blocks tilting backwards and moving on the same slide plane at the sole of the slumps (Selby, 1982). Rotational slides occur with a concave-upwards curved failure plane (rapture surface), passing through a thick and relatively homogeneous bed of clay or shale (Hutchinson, 1968).

4.5.3 *Slips Due to Natural Processes*

Most of these slips, also known as lateral spreading, occur in sensitive soils. This sensitivity is the maximum soil strength when undisturbed. Once slight movement occurs a large decrease in soil strength take place, thus a rapid failure commences. In undisturbed state these soils are well cemented, but once disturbed fissures develop and they are reduced to gravel sized aggregates.

4.5.4 *Block Glides*

These are slides in which material is undeformed as it moves over a planar slide plane. These are very small scale phenomena.

4.5.5 *Structural Slides*

These slides occur in bedrock or soils suffer shearing along a marked discontinuity, such as the junction of a stronger and a weaker bed. In consolidated materials the failure plane may be kept near the surface because of the general tendency within these beds for shear strength to increase rapidly with depth. Lateral movements of a large mass is generally on slip surface formed by bedding, cleavage, or joint plane, occupied by a filling of material having a low shear strength. The actual details of the distortion and the depth of the slide depend on the nature of the slide material, and the movement may include flow as well as sliding.

4.6 **Genesis**

Inasmuch as mountain massifs keep rising, and because rivers keep deepening the channels, all slopes comprising the relief have a tendency to become steeper with time, and mass movement is caused if the stability breaks down (Scheidegger, 1984). As stated by Ai (1985):

> "The ultimate cause of mass movements on slopes is, therefore, related to the operation of the normal geomorpological cycle."

4.6.1 *Generic Factors*

Causation factors of landslides though attributed mostly to natural processes, yet the most important part of it are related to unstable slopes and released cascading earth material, meanings, landslides pose severe hazard in mountainous terrain of varying slope angle.

An examination of world wide distribution of steep slopes indicates a strong material composition, and low margin of stability (Crozier, 1986). To examine the instability of slopes in a particular terrain, besides geochemical conditions, one must identify the history of prevailing slope environments (Fig. 4.5).

In buildup areas, mechanical tools used for construction cause drastic modification of the slope. Excavation, an instrument of anthropogenic activity, is capable of shifting a slope with stable height/gradient relationships towards its critical frontier within a short while. Modern technology has rapidly concentrated large weights on the ground translocating surface material. All these modifications of slope induce conditions of instability, ultimately causing failure.

The incidence of these hazards has been on increasing trend throughout the twentieth century (Coates, 1985), making it a part of double jeopardy syndrome. This is believed to be the result of increased anthropogenic interference with naturally unstable environments of highlands, through urbanization, transportation network, increased economic activities and a stock of many other examples.

These causes of landslide all depend on the threshold of earth material when the forces producing change are stronger than the forces of slope stability. Sliding of land takes place when slope materials are no longer able to resist the force of gravity. This decline

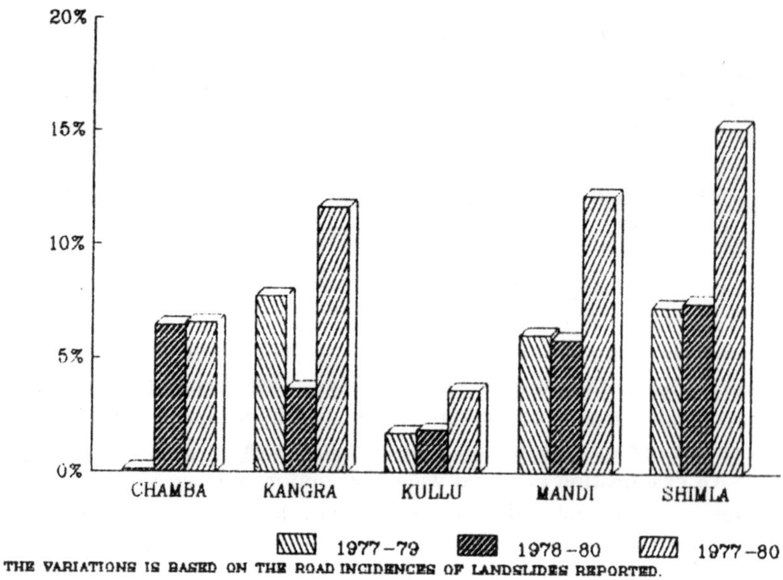

FIG. 4.5: Variation in Annual Debris Production Through Landslides (1977-80)

in shear resilience can be an aftermath of either external or internal causes. Internal causes lead to decrease in resilience (hence, failure), and are the eventualities of modifications in physical or chemical properties or water content of the material and can broadly be classified as follows:

4.6.1.1 *Physical Factors*

These are further divided into three which are as follows:

- Beds which decrease in shear resilience if water content increases (clay, shale) (e.g. when local water table is artificially increased in height by reservoir construction, or as a result of stress release [vertical and/or horizontal] following slope formation).

- Insignificant inner bond (e.g. consolidated clays, sand, porous organic matter).
- In bed rocks faults (e.g. bedding planes, joints foliation in schists, cleavages, brecciated zone, and preexisting shears).

4.6.1.2 *Chemical Factors*

These are further divided into two which are as follows:

- Weathering reduces effective cohesion, and to a lesser extent the angle of shearing resistance.
- Absorption of water leading to change in the fabric clays (e.g. loss of bonds between particles or the formation of fissures).

4.6.1.3 *Hydrolic Factors*

Higher ground water table as a result of increased precipitation or of anthropogenic interference (e.g. dam construction).

External factors are usually regarded as factors which enhance the shear stress on the slopes, often involving a form of disturbance that may be either natural or anthropogenic. However, a broad classification of these factors is illustrated as follows:

4.6.1.4 *Removal of Fundamental Support*

Following three factors are the eminent activating reasons:

- Undercutting by water (e.g. rivers), or ice.
- Washing out of granular material by seepage erosion.
- Anthropogenic excavation.

4.6.1.5 *Increased Loading*

Following two factors are the eminent activating reasons:

- Natural accumulations of water, snow, and taıus.
- Anthropogenic pressure (e.g. bank, edgeload, or building).

4.6.1.6 *Ephemeral Earth Tension*

Following two factors are the eminent activating reasons:

- Earthquakes.
- Frequent passing of traffic.

Thus while attempting to evaluate the causes of environmental anomaly, two main factors must be stressed, i.e. natural factors, those preparing the conditions for hazards and change, and anthropogenic. The eminent, but not sole cause of hazard occurrence. Thus, environmental stability especially in case of slopes, can be defined as the ratio of resilience to stress, and can be indicated as:

$$\textit{Stability} = \frac{\textit{Resilience}}{\textit{Stress}}$$

However, Petley (1984), discussed the stability of slopes, defining them as: [*Fs = Shear Strength / Shear Stress*], if $Fs > 1.0$ stability is likely, but if $Fs < 1.0$ instability exists. It is important to note that, to determine the values of Fs, onsite measurements of geotechnical properties of the slope material are needed.

Landslides do not occur in isolation, but also they are products of their environment and in response influence the environmental conditions (Crozier, 1986). Much of the literature on landslides has been event specific. As a result, what happens during a slide is reasonably adequately comprehended in strict physical terms.

4.6.2 *Probing Dilemma*

To probe into the combination of coercions invoking landslides, there is a requirement of the cognizance of both functional linkages between environment and slope stipulations, and ebullient of the environmental system. Among all the hazards associated to mountains, the mass movement is the most serious one. It can be termed as a part of natural geomorphological cycle responsible for landscape development, these movements usually can not be prevented (Scheidegger, 1988), but can be predicted and their effects mitigated (Pirazizy, 1991).

Hazards arising from mass movement are a common problem in mountain regions, particularly in a temperate humid climatic conditions. However, the ultimate cause of all mass movements in the general pattern of landscape development. A landscape is an open system in a dynamic state, which has remote chances of achieving equilibrium. Studies reveal that tectonic activities is the main phenomena in mass movements (Scheidegger, 1988), though at first glimpse many confuse it to be entirely caused by external agents.

4.6.2.1 *Generic Surroundings*

This includes evaluation of topography, geology, climate, information on existing roads, and alignment of roads to be constructed.

4.6.2.2 *Intention of Construction*

This includes estimations regarding the purpose and serviceability of roads considering area and its population and volume of traffic to be accommodated during the first few years.

4.6.2.3 *Technical Aspects*

This includes estimations on road length, width, type, cuts, heights, and slope.

4.6.2.4 *Environmental Stability and Suitability*

This includes hazard zoning, evaluation of slope stability, vegetative cover, suitability, and mass wasting risk assessment.

4.6.2.5 *Environmental Impact*

This includes appraisal of impacts on slope stability, hydrological conditions, floristic ecosystems, and landuses.

4.6.2.6 *Measures for Reduction in Impact*

This includes all technical and socio-economic concerns and techniques used in remedial and immediate responses.

In view of all environmental problems related to road construction the environmental impact assessment before enacting any project is must.

4.7 **Dam Construction and Environment**

Power generation and accelerated industrial growth have been the basic reasons for establishment of dams in increasing numbers, which have proven to be incompatible with regional environments especially in higher elevations. In proceeding sub-sections it will be observed that dams and their reservoirs have been/are potential agents of environmental degradation subsequent to consequent hazards and respective risks. As stated by Lagler (1971):

> "Secondary costs of hydrodams take the form of environmental alteration, destruction, or obliteration and involve human, water,

land, and living terrestrial and aquatic resources . . . ignoring known concepts of (environment) . . ."

Reservoirs can be interpreted as anthropogenic counterforce, which change unstable riverine ecosystems into stabler lacustrine ecosystems, arresting the discharge completely, which under less shear geolithological resistance of the zone can be disastrous. Moreover, stabilized water levels of reservoirs affect the energy flow regime, thus, reducing turbidity through depositing action, hence, intensified sedimentation. The first correlation between reservoirs and earthquakes was deduced in Greece in 1931 (Coates, 1987). One of the major causes of this correlation in India was in 1967, tremors of Koyana reservoir, which caused killing of 200 persons and injury of another 1500, and Bombay, 230 km. away, was shaken (Gupta, 1976). The mechanism is postulated to be related to shear fracturing of rocks. In areas where rocks are under initial shear stress affiliated to a fracture plane, the increased pore pressure by water movement may overcome the frictional resisting force accompanied by shear failure. The environmental changes associated with dam construction are many, and the important of all are those causing hazards with unprecedented risks empirical evidences of which are listed as follows:

4.7.1 *Kol Dam*

This reservoir is situated between Dehar and Tatapani in Shimla district, is facing the problems of tightness stability of rim, and siltation due to landslides. It is made on a lithological structure of Shalis, Basantpurs, young intrusives, and Quaternary and Recent sediments of terraces. The rocks are highly fractured and jointed which make the area highly landslide prone (Jalote and Mithal, 1971). The Dolomites and Limestones belonging to Shalis in the vicinity of Kol dam, show local sinking in the form of depressions that are not aligned with structural fractures of the rocks but in the interruption of slopes. One of these depressions developed to 259m. on the hill of Sundernagar (Beas-Sutlej Link Project), with several smaller depressions between Sundernagar hills till Kol dam site.

4.7.2 *Baira-Baledh-Siul Link Project*

This hydropower scheme is located in Siul basin (Chamba district).

The terrain forms a part of topographic relief dominated by slips and other mass movements, and is characterized by deep gorge section and valley asymmetry with joint controlled rivers. The slopes consist of rock escarpments interspersed with talus/scree accumulations (Gosh, 1980). Geographically the area is composed of Phyllitic Slates, Slaty Quartzites, Pebbly Slates, Carbonaceous Slates/Phyllites, with interbedded Limestone and Dolomite, and low to steep dips in NW-SE trend across the valley. The different rock units have been thrown into a major syncline (Kalhol syncline), and fold axis trend of NW-SE. Two major faults, Dhar fault and Manjir fault, have developed along the core of fold axis and the project area lies on the north-easterly limb of the fold. From the description of terrain parameters, structural discontinuity planes and slip occurrence through the weight of spillways, it is evident that the rock slopes are slide prone, and project must have been located after careful consideration of stability. The problem of instability has increased many folds, because besides tectonic residual stress, the forces resulting from other stresses such as loading, blasting and earthquake have not been considered, as the rock slopes of the area are unstable with a low safety factor, and slide direction is along divisional planes against steep slopes.

4.7.3 *Pandoh Dam*

The environmental problem this dam is causing is in the right side of the reservoir, which lies on highly fragmented and jointed Chail Phyllites formation. This section of reservoir is drifting through block-gliding, caused by gravity, having started a wide open crack on the supporting hills . The spur forming this abutment is under rigorous base erosion by the Beas, and joint patterns are facing a depositing process, making the block-glide to be deep seated. The slide has a great potential, which can be observed in its having already deflected the slide mass from the alignment of dam down stream, hence, filling the reservoir with draining material.

4.7.4 *Pong Dam*

The main problem in this reservoir lies in its tunnel intake area, which is located on geologically weak structure of sandstone, claystone, and shale in an alternative sequence. This tunnel intake is situated in the left abutment of the reservoir and due to plasticity of geological structures (clay and shale connection bands), the area has wended slides in the form of block and planar drifting. The beams dug against these plastic geological bands have developed low shear capability, the head of slides is overloaded, and sedimentary sequences are void of concrete shafts to add shear resistance to common planar movements.

4.7.5 *Baira Dam*

This dam is under construction on Siul river, a right bank tributary of Ravi river. The site of construction is in the close vicinity of landslide area, which is composed of geologically highly jointed Phyllites of Chamba formations. Strong base erosion of these steep slopes of joint oriented rocks, due to high degree of slope and instability of the joints, is an active site of falls and slides in planar pattern, which move as debris slides. Until the present these movements have twice completely blocked the Siul river's coarse. The basic reason of such instability is the lack of deep subsurface drainage of the site.

4.7.6 *Giri Hydel Project*

These (series of projects) are sited on weak Quaternary Sandstone with bands of claystone along low angle dips. The cuts of the power house are parallel to joints (claystones), consequently resulting in planar failures, hence, large scale silt accumulation in the excavations. There are no strong retention walls; excavations, meant to locate the power house equipments, are of open type which add many folds to the instability of the area.

4.8 **Postlude**

Though the developmental activities create many localized environmental problems, yet the interaction between these problems gives them a regional dimension. Thus, the basic purpose of policies concerning environment and development must be concerned to enhance life quality and satisfaction of human needs. As Tolba (1982) owns it:

"We may now look upon environment as the stock of physical or

> social resources available at a given time for the satisfaction of human needs, and upon development as a process . . . (aimed to increase) human well being".

Despite the massive extent of literature and research work available, far less is known about complexity of environmental system prevailing in highlands. Besides, considering the natural rates of deterioration and its accelerated counterpart, assessing consequences of change is quite intricate. Surface disturbances of physical environment are capable of producing increased erosion which in other words means greater concentration of sediment load in streams. UNESCO (1982), has an extensive documentation of the effects of different agricultural activities in increment of sediment load of the rivers. Yet, greatest contribution to increase in sediment load is an attribute of construction ventures. In a study conducted by Wolman and Schick (1967), the amount of suspended sediment concentration was recorded for streams passing through construction sites was 3000–150000 ml/l, while the corresponding figures for agricultural watershed was 2000 ml/l.

Moreover, the mean figures for rural catchments are about 500–600 ml/l. This study concludes that, prior to the development of agriculture, mean figure was 50 ml/l, rising to 600 ml/l during intensive farming, and further increasing to 2000 ml/l during the period of construction activities. Landslides alike other high energy processes, acquire hazard status, once anthropogenic environment is vulnerable. Main factors increasing the vulnerability of any area, in highlands, to landslide damages are urbanization, increasing reliance on transportation, and service systems. These factors are consequences of anthropogenic technical capability to modify the physical environment. In most of the instances, landscape anomaly has enhanced land crust sensitivity, increasing landslide frequency, examples of which are listed in antecedent subsections. Landslides, if not extensive, have a short duration impacts. However, in inherently unstable highland areas of the zone, frequent and extensive slides are common phenomena being quite lengthy and severe. The effect of debris produced through slides reaching the streams is mostly magnitude related, i.e., capability of sediment to be transported and its stability towards the action of fluvial activities (Grant, et al., 1987). Some slides may develop into the swift torrents completely striping vegetative cover, surface soils, and channel sediments from large distances of channel.

CHAPTER FIVE

ASSESSMENT OF HAZARDS

5.1 Prelude

One of the most energetic areas of research on anthropo-environmental interactive systems is the hazard appraisal. The initial work on this aspect has lain with geographers, and in particular environmental geography. Research on anthropogenic environment of geophysical events has both practical and academic importance. The gravity is manifested by the dilemma of the eventualities emanating to be the result of technological progress, rather than being averted by it. This aspect results in perplexity regarding the study of hazards. The actual source of this enigma lies in the weak distinction between natural and anthropogenic hazards. Kirkby (1973), attempted to discuss these differences as follows:

> "Generally speaking natural hazards tend to operate from the physical environment on man; man-made hazards are the effects of man's activities on man through the medium of the physical environment. . . ."

Natural hazards, is a term indicating action of abrupt modifying natural events, such as floods, droughts, earthquakes, landslides, and the like, while anthropogenic hazards are indicated by the term *technological hazards or environmental risk* (O'Riordan, 1986). The routine problem of terminology has impelled the division in the whole encompassing term *Environmental Hazards*, into *Natural hazards* and *Technological hazards* or *Environmental risk*, terms, to facilitate the assessment in proceeding parts (Pirazizy, 1991).

A hazard by definition is an expected damage, injury, or death to people and property, subsequent to general distress, and economic loss. As will be elaborately indicated further in the present chapter, a natural hazard is considered a problem if occurring in built areas, while people are not aware of its characteristics in order to cope with it. Hazard and risk research is an important sub-area of investigation in environmental geography, and an consequential element of the anthropo-environmental interactive systems, which is central to geographical studies. In anthropo-environmental interactive system, due to erroneous planning of resource utilization, many negative eventualities of hazards and consequent risks have resulted.

Natural hazards pose serious problems in the zone. It is actually very difficult, within the limited research ability of the individual, to compare the relative effects of different natural hazards, because of differentiation of imposed situations. Natural hazards problems exist over a wide range of scale and aspect. Each hazard is peculiar in its situation. The relative magnitude of natural hazards depends on the actual hazard process itself, beside other factors. In terms of hazard, the variations between fast and slow hazard can be analyzed in terms of potential avoidance velocity. The basic dilemma in hazard studies still remains in the ambiguous parameters used to differentiate between natural and anthropogenic hazards. In the present book the choice of approach to the problem of hazard evaluation depends on following points:

- Location,
- Impost,
- Duration;
- Accessibility.

The need for consideration of the above four aspects was with the regard that the hazard map construction in the zone is one of the very few pioneering attempts towards a hazard risk map construction. Secondly, the gathering of greater information densities were beyond the budget and time allocation to the present work, thus, the optimum method in view of constraints and cost/benefit ratio were selected.

5.2 Empirical Evidences Through Analytical Persuits

Eventhough the hazards in the zone have been the generating source of various problems, yet no map of the geomorphology, hazards, and associated risks of the zone has been prepared. The reason being that, the Indian Himalaya in general and its Northwestern portion specifically, has been a case of slight. The present analysis has used a mapping process, trying to present the main characteristics of the zone with empirical evidences from Shimla district. To attain the present goal map of geomorphic hazards and risks has, been discussed.

Shimla district is situated in the eastern portion of temperate Himachal Himalaya (30°45" and 31°44" N 77° and 78°19" E). The main river systems draining the district are Sutlej and Giri the major tributaries of Indus. The district is highly ragged and dissected, extending its borders from the cold desert of Kinnaur in east to the lower Himalaya in south. The predominant particularities of the area could be broadly listed as follows, using the division method adopted by Kienholz (1984), Zimmermann (1986), and Fort (1987).

Topographically the area forms highly steep gradient ranging from the lowest 650m. to the highest 5800m. (Fig. 5.1). This aspect in return has resulted in rapid bioclimatic gradient to aridity i.e. from southwest to northeast and from the glacial heights to the valley floors. Rainfall and rainfall changes for the year 1986 (monthly basis), and 1976-86 (decadal basis). The rainfall data was collected from the Directorate of Land Records for the Shimla Station. The distribution of rainfall is typical of monsoon Himalayan station; maximum amounts during four months of summer, the minimum by the late autumn and early winter, followed by a progressive increase of precipitation due to winter *Mediterranean type* depressions (Fig. 5.2). Above 4000m. altitude the snow cover remains for about six months and there is a spatio-temporal spread in between 3500–3000m. In places of avalanche occurrence there is a limited scale prevalence of ice sheets. During mid March to June there is moisture sufficient conditions. Frost, which is an active agent in rock disintegrating, occurs almost through out the year. Geologically the area falls in lesser Himalaya, which is marked by a change in lithostratigraphy and abrupt break between to lithotectonic zones. The zone is highly folded and faulted. Lithologically the area is weak and contrasted from sedimentary rocks, such as thick accumulation of clayey debris overlain by chail phyllites, which are overburdening interface and deflecting north wards. Presently active process of orogeny and dynamic process of cataclysm, besides causing local bedrock causing local bedrock fracturing, resulting in leviathan mass erosion. Population density of approximately 100 persons/km^2 is fast increasing. Most of settlement concentration is confined to the low terraces of

Sutlej, Giri, Pabar, and their tributaries. Irrigated agriculture lands in the valley bottoms and immediately adjacent terraces, dominantly cultivated by maize and wheat, are practiced. Most of the economic benefits of the area comes from off farm activities like livestock, mostly sheep and goat, and forestry. Recent years have experienced increasing dependence of local economy on orcharding, with side effects on silver fir and spruce forests as a result of attaining prerequisite of packing cases, tourism, and administrative activities (Pirazizy, 1991) (Plate 5.1).

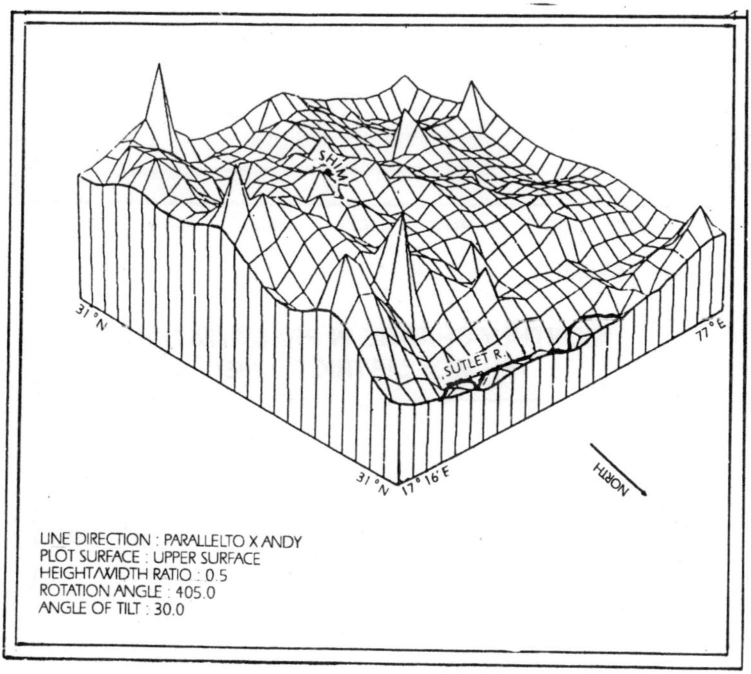

FIG. 5.1: Surface Peresentation Shimla District

Natural hazards modify relief conditions as well as topographic gradients, are primarily decided by the type of environment prevailing spatiotemporally at any place. These local environments are made up of different sustaining components with specific resilience and inertia, well adapted to the local conditions. The anthropogenic choice factor to hazards is formed by the way hazards are defined.

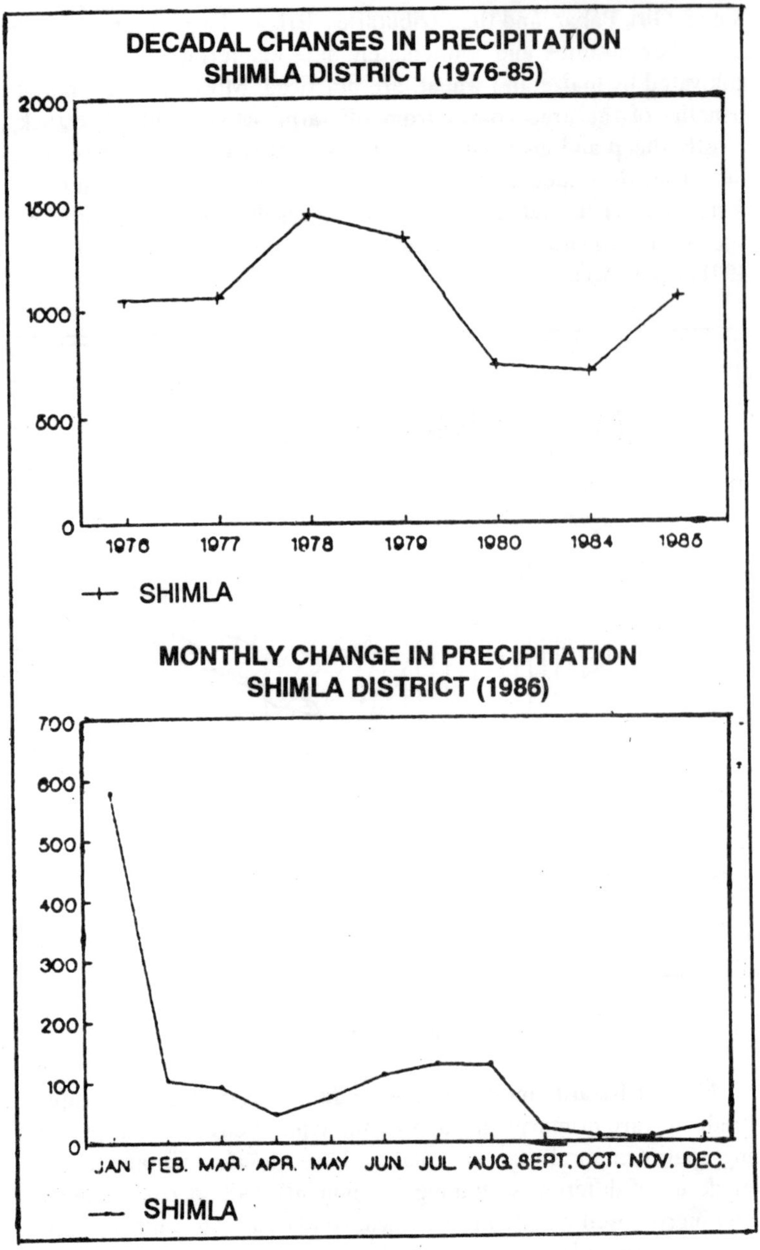

FIG. 5.2: Changes in Precipitation Pattern: A Comparative Analysis of Hazards Area

PLATE 5.1: Rockfall near granite structure, caused as a result of many fractures, which reading yield to mass wasting. Usually road beds and cut slopes are at a very high angle of about 25–30 degree. Seasonal saturation and meager vegetation cover make these slopes vulnerable to failure ensued by intense note the high degree of slope.
(a) Between Kufri — Fagu (1990)

PLATE 5.1: (b) Between Rampur — Narkhanda (1990)

For instance, drought is a hazard as defined by a researcher, but rainfall variations are environmental dimensions according to a farmer in Temperate zone Himachal Pradesh. Thus, the individual or community response of farmers to spatiotemporal rainfall patterns are more complex than a undimensional set of adjustments to drought. A glimpse on the ethnography of the zone indicates that in its prevalent traditional culture, concern to classify hazards by their different effects is less acute than the need to identify the chain of cause and effects. Till the genesis of effects are not found no adjustment can be adopted by the locals. Moreover, people's taxonomies are also different, to differentiate between anthropogenic and natural hazards, from the established research ones. Thus, spatial extent of maladaptive hazards, may be adaptive in the local context of its risks (Whyte, 1986). Thus, any change in physical configuration of any area directly affects its environmental sustainability (Pirazizy, 1991).

5.3 **Historical Purview of Hazards**

Till the early 1900–, peruses of natural disasters were merely descriptive in nature. Scientific scrutiny of physical causes responsible for such disasters, were then established quite fairly. Yet, Their relation to society and anthropogenic perception were but little. Primarily, the main emphasis on any research relating to hazards was based on anthropogenic hazards (O'Riordan, 1981). A gradual realization was being made on the relevance of these findings with natural hazards. However, by 1950–, the time of establishment of a Committee on Disaster Research by National Academy of Sciences U.S.A (1953), and a decade later, the parallel research of social responses to natural and anthropogenic hazards, attributed an academic respectability to hazard research.

By 1970, a paper was published by Natural Hazard Research, which aimed to create a format for data collection on, physiography, geomorphology, human use system, frequency and magnitude of hazards, estimates of damage, and list of hazard adjustments of the areas under research then. The research continued for improved techniques and more careful research, examples of such researches are evident in scenarios proposed by Erikson (1974), and many others. By 1983, Burton, Kates, and White summed up the preceding works in their collective work, explaining hazard syndrome as a research agenda for future more detailed works yet to come, indicating that:

"The theory holds that in using physical resources people engage in behavior that combines adaptation to extreme events with both purposeful and incidental adjustment . . . the many and varied coping actions can be grouped into . . . loss absorption, loss acceptance, loss reduction, radical changes . . ."

5.4 Geomorphic Appraisal

Prior to development of a perspective on hazards in Himalaya, mountain geomorphology must be considered as a broader issue along with the evolutionary trend of it during the last few decades. Important characteristics in the geomorphic nature of highlands include significant structural influence in the landscape. Evolution of geomorphology evolves the broad concept of landscape development. Majority of highland terrains show evidences of ancient erosion. Any landscape is evolved through geomorphic processes, the dominant of which according to Rapp (1960), are silt transportation in runoff, landslides, avalanches, rockfalls, glaciation, and talus creep. These geomorphic processes are characteristics of all highlands, yet due to variation in natural parameters such as climatic conditions and vegetation cover, these processes are spatially variegated. The resulting landscape possess high relief and greatly varying potential energy, which is reflected in the history of recent erosion by natural hazards, and high accelerated erosion potential at the vicinity of built-up environments (Barsch and Caine, 1984). However, anthropogenic interferences alter the pace, frequency, and magnitude of these modifying processes. Ranking of geomorphic processes, or natural hazards is different due to differences in pace and magnitude (e.g. slow processes such as glaciation, and fast processes such as rockfalls). Inaccessibility of Himalaya makes the ranking of these processes further difficult (Pirazizy, 1991).

5.4.1 *Structure of Northwestern Himalaya*

A very important perspective necessary for understanding geomorphic processes is the rate of Himalayan orogeny. Zeitler et al, (1982) estimated a rate of 1 mm/yr. uplift for greater Himalaya and 9 mm/yr. for the Nanga Parbat region. These facts along with more evidences approve of massive erosion and regional denudation. These crude rates of denudation hardly equal those of uplift estimates, thus, Himalayan system has been an extremely dynamic land mass over the last 10,000 years.

Physiographically the entire zone can be broadly divided into four west northwest to east southeast strike oriented units (Ives and Messerli, 1989). A transition belt joins the southern flanks of Greater Himalaya with Middle Himalaya, which is densely populated. Middle Himalaya, the transition zone and the Lesser Himalaya form a tripartite units which are under Upper Montane Forest (2900–4000m.). Below this zone is the belt of intensive agriculture. Lithologically this zone is made up of sedimentary, metamorphic, and igneous rocks, with extensive areal distribution of Phyllites and Schists which are highly disintegrated and in presence of steep slopes have become strongly susceptible to erosion (Pirazizy, 1991).

5.5 Risk Assessment Analysis

Environmental risks in the zone, as an example, are almost all built area specific, and are of uncertain magnitude, frequency, and cause. Natural hazard may be defined as extreme geophysical event (Oliver, 1979), immensely exaggerated according to normal occurrences both magnitudewise and frequencywise. Events such as floods, blizzards, tornadoes, earthquakes, volcanic activity, landslides, avalanches, rockfalls, mass movement, soil erosion, and drought amongst many other are considered as natural hazards. Natural hazards, however, are the result of complex interaction between various natural systems as stated by White (1974a):

> "Where there perfectly accurate predictions of what would occur and when . . . in the intricate web of atmospheric, hydrologic, and biological systems, there would be no hazard."

Hazard is believed to be a natural event expected to occur in a definite spacio-temporal definition. Hazards are risk oriented as far as anthropogenic environments are concerned or *when things valuable to people are at stake* (Dibble et al, 1984). The analysis of risk involves estimation, identification and assessment of–

- Nature and degree of hazard;
- Damage and its socio-economic eventualities; and
- Degree of probability of occurrence (Petak and Anderson, 1982).

It is crucial for hazard risk assessment purposes to know whether a hazard is active or dormant. There are various criteria which permit distinction between these two states, based on surface evidences (Table 5.1).

TABLE 5.1: Distinguishing between Active and Dormant Hazards

Active
– Scraps, terraces and crevices with sharp edges, – Crevices and depressions without secondary infilling, – Secondary mass movement on surfaces, – Surface of rapture and marginal shear planes show fresh slacken sides and striations, – Fresh fractural surfaces on blocks, deranged drainage system; many ponds and undrained depressions, – Pressure ridges in contact with slide margin, – No soil development on exposed surface of rapture, – Presence of fast growing vegetation, – Distinct vegetation differences on and off slide, – Tilted trees with no new vertical growth, – No new supportive, secondary tissue on trunks.
Dormant
– Scarps, terraces, and crevices with rounded edges, – Crevices and depressions unfilled with secondary deposits, – No secondary mass movement on scrap faces, – Surface of rapture and marginal shear planes show old or no slacken sides and striations, – Weathering on fractured surface of blocks, – Integrated drainage system, – Marginal fissures and abandoned levees, – Soil development on exposed surface of rapture, – Presence of slow growing vegetation, – No distinction between on and off slide, – Tilted trees with new vertical growth above inclined trunk, – New supportive, secondary tissue on trunks.

However, in conditions as complex as temperate Himalaya, with multitudinous differences in altitude, micro-climatic conditions, aspect, and anthropogenic environments, such a distinction is rarely simple, and arbitrary decisions often have to be made. For example, often one comes across the question that a landslide which happened say ten years ago is an active slide? The term perhaps best reserved for slope failures that have shown movements in the recent past are expected to show movements in near future again. Other slide hazards may be recent rather than active and can be considered to be historical in that they can be assigned to some past period. For the present study this precise form of such classification does not matter too much, what really matters, however, is that hazards are recognized for what they are. In advertent mismanagement, such as an excavation into the toe of an inactive hazard, may cause it to become active, at times with devastating aftermath. For better understanding of active and dormant hazards, based on Crozier (1984), following features have been enumerated for distinguishing between these two states.

5.6 **Delimiting the Area**

Complexity of highland hazards is an attribute of their collective action. Thus, for the assessment purpose "Unit Area Method" (Kienholz, 1978), which compiles areas of homogeneous hazard occurrence potential, is used. This method of delimitation is directly based on correlating obvious phenomena. The priority of mapping in this method is given to the hazard which has larger spatial coverage, in case of collective action of hazards. Therefore, the zone was delimited according to prevalence of hazards and its inherent instability, which facilitates the evaluation of causal parameters. Finally the results and information obtained were assembled into a hazard map (Fig. 5.3).

5.6.1 *The Geomorphic Features Map*

Natural erosion is the widespread phenomenon throughout the Shimla district, though at present greatly accelerated. The geomorphology of the area displays two prominent features comprising of contrast between both slow and rapid processes, which are generally local phenomena. Both these processes which could be exemplified as mud flow (Kufri-Fagu belt), rockfalls (Rampur-Sarahan belt), and debris flow (Shimla-Dhalli, Shimla-Tatapani belts) result in transfer of large quantities of substance. High inclination gradient in topographic features is the basic cause involved in such geomorphic activities. Geographic location, and geological conformations (bands, joints, faults and folds), and composition (Shales, Schists, etc.) are among the other causal factors.

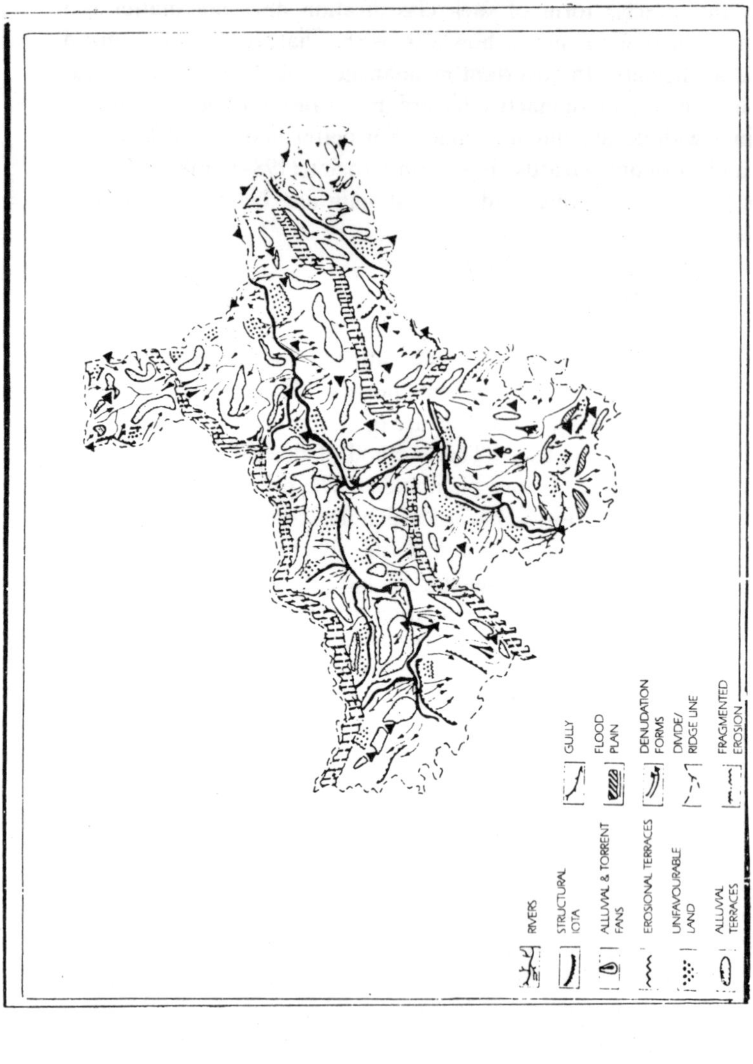

FIG. 5.3: Shimla District: Geomorphological Landscape

Anthropogenic impact can be observed along the irrigated terraces, roads, settlements, pastures. These impacts have resulted in subsequent top soil loss through slumps and earth flows on favourable substrata of Shale, Litt, Schists, and Lacustrine deposits, along the irrigated fields and terraces. Anthropogenic activities through deforestation have created daunted forest land cover over 40–50 per cent of the district. The deforested lands have been brought under agricultural, grazing, and construction landuse types, hence, topographic modification. These modifications of local relief to extensive terrace construction from top to bottom of slopes is evident on many aspects, especially along the river valleys. Application of irrigation techniques, has directly affected water resource conditions. Thus, in general, most of the Shimla district can be considered as the anthropogenically modified landscape (Fig. 5.3).

5.6.2 *Basic Geomorphic Processes and Consequent Hazards*

Like most areas of the earth's surface, Himalayan highlands experience a variety of natural hazards, which do not justify the western definition of hazard events. Geophysical experiments have proven that the height of present time Himalaya is the same as what it was about a million years ago, with undiminished relief energy between crest line and plains. This proves the insignificant degradation contribution of modern human beings during the last few decades, when compared to natural balance created by dynamic exogenic and endogenic processes.

When massive scale of altitude, great deal of variations in climate, vegetation cover, and micro level topography is considered, then, attaining the actual rates of change and active geomorphic as well as orogenic processes acquires a very difficult dimension, if not impossible with the limited time and facilities devoted to the present task. However, despite the impeding difficulties, if the recent geological past is considered, then some valuable contribution to geomorphic evaluation could be made. The present task could be achieved if the assumptions that anthropogenic ventures are the primary causes of mountain instability due to wrong application of technological knowledge to sensitive highlands is investigated deeply, some of the major geomorphic processes are listed as follows:

5.6.2.1 *Soil Erosion*

Parts of Sutlej and Giri rivers draining the Shimla district, are sustained by multitude of tributaries situated in 141 sub-watersheds in an area of about 260000 hectare. From these sub-watersheds, seven major sub-watersheds have been selected, for reconnaissance, with an area of about 19000 hectare. Out of the total selected sub-watershed units about 53.58 per cent are under very high erosion intensity category (Table 5.2). High and medium categories account for 34.30 and 12.12 per cent of the area respectively. Prevailingly, about 57.14 per cent of sub-watersheds in the appraisal units are of very high erosion intensity category. Figures for high and medium categories are 28.57 and 14.28 per cent respectively (Table 5.3) (Fig. 5.4).

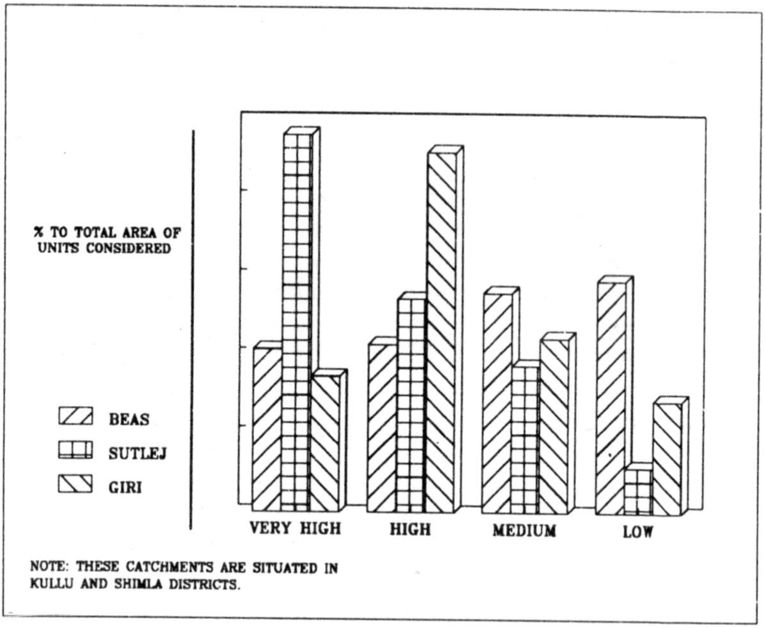

FIG. 5.4: Erosion Intensity

As inferred earlier, the very high erosion intensity units in the district requisite an contiguous conservation-cum-treatment to intrinsically curtail already high silt load of the rivers.

Table 5.4, elucidates the proportionate distribution of erosion types selected for probing, assisting in extensive characterization of the catchment and the operating landuses (Fig. 5.5).

TABLE 5.2: Erosion Characteristics

Sutlej and Giri Catchment	Area (ha)	Erosion Category Range	Silt Road Index tonnes	% Area to Total Catchment
Seoni	3222	Very High	>1300	17.36
Shimla	3944	High	1200-1300	21.25
Narkhanda	3924	Very High	>1300	21.14
Govind Sagar	5224	Very High	>1300	28.14
Chaupal	2250	Medium	1000-1200	12.12

TABLE 5.3 : Spatial Share of Erosion Intensity Category

Erosion Category	Silt Load Index Range	Area (ha)	% Area to Total Catchment	Number of Sub-water-sheds
Very High	>1300	9946	53.58	4
High	1200-1300	6368	34.30	2
Medium	1000-1200	2250	12.12	1
Total		18564	100.00	7

TABLE 5.4 : Percentage Area Under Various Types of Erosion

Sutlej and Giri Catchments	(I)	(II)	(III)	(IV)	(V)	No Erosion
Seoni	10.17	8.53	57.84	7.42	–	16.04
Shimla	–	5.17	26.27	63.18	5.38	0.00
Narkhanda	0.51	56.88	11.72	29.15	1.73	0.01
Govind Sagar	8.57	–	91.42	–	–	0.01
Chaupal	–	16.70	44.40	38.90	–	0.00

This type of geomorphic process occurs in both entirely natural and anthropogenic (accelerated erosion) modified environments. Some gross figures indicate that main river catchments in Himalaya produce about 14.6 million tonnes of silt annually (Naravana, 1987), while, another figures by Singh and Gupta (1982), indicate 75 million m^3 (about 1.7 mm/yr. denudation rate). These figures, however, when considered on regional level depict great accuracy (Pirazizy, 1991).

5.6.2.2 *Mass Wasting*

In general the down slope movement of fractured and weathered bedrocks, over which soil forms as a product of natural process of weathering, is termed as mass wasting. Mass wasting can be either slow process of creep, solifluction, etc., which are continually active under the influence of gravity with a rate ranging from a few millimeters to a centimeter, as well as fast moving interrupted events such as rockfalls, mud-flows, landslides, avalanches, and rock slides. Calculations show a denudation rate of 2.5 mm/yr (Ramsay, 1985), on an average due to landslides only.

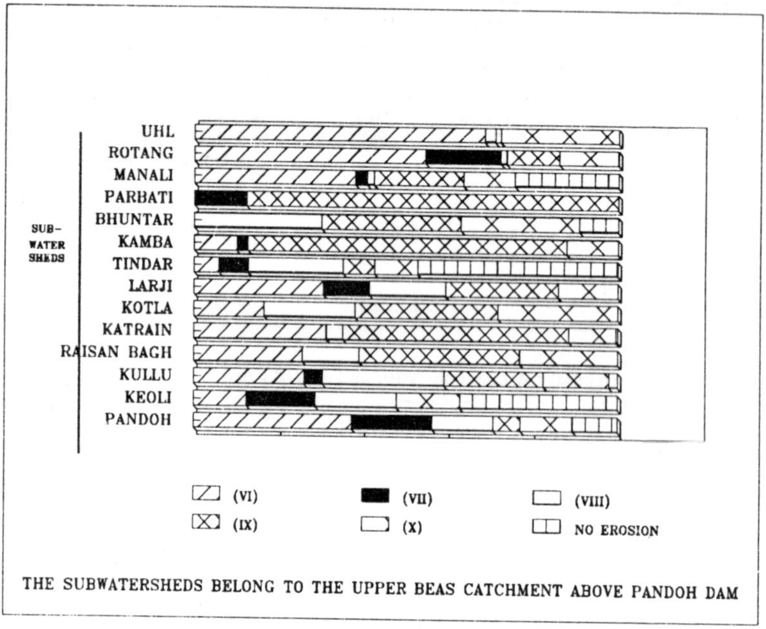

FIG. 5.5: Proportionate Area Under Different Erosion Categories

5.6.2.3 *Rock Falls*

Stream erosion, percolation of irrigation water through the lithological joints, deep weathering, anthropogenic cuts on slopes, and instability of lithological blocks are major causes of rockfalls in the zone. South facing slopes towards northwest, i.e. Shimla-Mandi, Shimla-Kullu, and northeast, i.e. Shimla-Theog, Shimla-Narkanda, Rampur-Sarahan, are the zonal belts of major rockfall in the district. Only areas with lacustrine deposits do not experience rockfalls at the bottoms (Plate 5.2).

PLATE 5.2: The generated material through failure and landslides in the zone is untransported weathered bed rock, consisting of sand, silt, varying proportion of gravels and boulders. Most of these structural changes occur within the 25 cm. of the surface layer, where anthropogenic impact is prominent. The nature of the soil in the zone emphasizes the significant role of water in commencing slope, failure. Rainfall in the important parameter, which needs detailed consideration, but little information is available to relate specific landslide activity in the zone with precipitation.
(a) Between Kullu — Manali (1990)

PLATE 5.2: (b) Between Shimla — Kufri (1990)

5.6.2.4 *Landslides*

Landslides of the area are generally, of shallow depth (>2.5m. of top layer), and are common features in hollows and valley heads. Following the initial failure, most of these slides disintegrate to become debris flow or flow-slides, which most of the time do not transport the failed material far from the site of failure. Within the failure scar, further development of slab failure of its walls take place. The debris accumulated at the base of failure, may be removed either by, periodical failures as earth-flows, action of running water, or, through anthropogenic interference. However, for the zone, the term landslide is used to depict all forms of primary and secondary mass movements such as, debris-flows and earth-flows, which are common all over (Plate 5.3).

5.6.2.5 *Water Erosion*

By causing various depths of vertical cutting, the commence of this process is through simple gully erosion. In the present inquiry E symbol and its forms subscribed by Arabic numbers, signify various types of water erosion. On the weathered Phyllites of Shimla district, there are many evidences of water flow below the surface, exposed through roof-collapse, producing steep-wall trenches. Besides ridges and peaks, the unirrigated terraces are areas of evident danger, where the slope degree is between 15 to 30.

5.7 **Construction of Base Map**

This is a map of all essential information of the area, and forms the foundation of the actual hazard map. The basic information for preparation of this map is collected from maps of landuse, geomorphology, geology, precipitation, soil, drainage, erosion, and precipitation. Essential informations contained on map are lithological boundaries, the drainage characteristics, physiography and the like.

5.7.1 *Mountain Hazards Map*

The steps needed in direction of reducing eventualities of hazards, stress on preparation of hazard maps based on geological and geomorphological information, recording of past hazards, frequency, magnitude and extent of the present hazards. Hazard maps provide estimates of the magnitude of risk, probability of frequency of

PLATE 5.3: Geologically Kalka — Rampur is made up of shali, limestone, siltstone, and orthoqurtzites of varying lithology. Usually the roads exhibit the most representing perimeters of degradation hazards. Notice the accumulation of road debris and screes of road drived debris down slope of the road bed.
(a) Sanjuali

PLATE 5.3: (b) Kalka — Rampur

occurrence relative severity, immediate response, and remedial response. In appraisal of risk associated with hazards following broad factors must be kept in consideration:

5.7.1.1 *Relative Risk*

This factor is the evaluation of risks likely to be encountered in different situations. The method used is arbitrary scaling of each situation and comparing it with another.

5.7.1.2 *Risk Probability*

This factor examines the damage resulting in terms of temporal frequency of hazard. This factor is calculated by following formula (Bolt et al, 1975),

$100(1-e^{-t/m})$,

m = mean return period,

t = future period of 100 years.

5.7.1.3 *Immediate Response*

The provision of reconstruction and remedial measures are taken care in this factor.

5.7.1.4 *Remedy Response*

The quantum of efforts for reconstruction determined by the urgency and affinity of anthropogenic environment is the highlight of this factor.

Thus as mentioned earlier in planning for hazards a sum of above factors are must, without which the evaluation procedure is incomplete and ambiguous.

Hazard map is multifunctional (Fig. 5.6), depicting, spatial extension of stable belts, and distribution and types of hazards. Basically the main aim of the method adopted is to classify different areas with gradients, according to the magnitude of potential stability, and instability. In return, this method facilitates the identification of areas with equal degree of the hazard risk (Fig. 5.7). The first step in construction of hazard and risk map was the (Fig. 5.7), which depicts the degree of hazard and risks. Due to limited research facilities, four numbers (i.e. 1, 2, 3 and 4), were selected depicting the degree of hazard occurrence. The areas of each degree were shaded with suitable technique. However, if the hazard map is compared with the buildup area map, the inherent danger of hazards and their degree of severity could be assessed. The degrees selected are as follows:

(VI) (Very Severe) : Indicates areas where failures have already occurred and are more likely to occur. Irreversible slope and structural destruction has occurred inhibiting remedial measures. This category covers about 21 per cent of the total geographical area of the district.
(III) (Severe) : Depicts areas where distress was observed, local sporadic failures have occurred, and the influence of erosion limits the possibility of restoration. The area under this category constitutes about 27 per cent of total geographical area of the district..
(II) (Moderate) : Indicates areas with occurrence of moderate hazards, depicting signs of moderate surface erosion, creep, etc., and capability of being restored within a long period. Moreover, areas remote from anthropogenic environment, even with severe degree, have been classified under this type. About 28 per cent of total geographical area of the district falls under this category.
(I) (Slow) : Depicts the immaturity of hazards with limited superficial damages. In other words, the more or less stabler slopes are governed by the parameters which favour the stability. The total area under this category covers about 15 per cent of total geographical area of the district.

For instance, Sutlej and Giri sub-watersheds, covering an area of about 260,000 ha., being rugged with altitudes between 650-5800m. are congenial for high runoff and silt yield. About 52 per cent of the catchments are under IV category (Very Severe) hazard risk, and are mostly confined to steep tributaries and settlement belts, experiencing unpredictable debris flows. About 25 per cent area is under III (Severe) category, which mostly experience localized hazards such as debris flows induced by severe slumps and gullying upstream. About 14 per cent area in these catchments is under II (Moderate) category hazard risk, though, they at times are very severe, yet due to occurrence in high altitude remote areas, with rare or absent anthropogenic activities, are rates so. And finally about nine per cent of the area is under I (Low) risk category, which are restricted to flat terraces under cultivation, or partly covered by forests, or by grasslands, or gentle slopes (0–5%), with slow and superficial processes of limited extent.

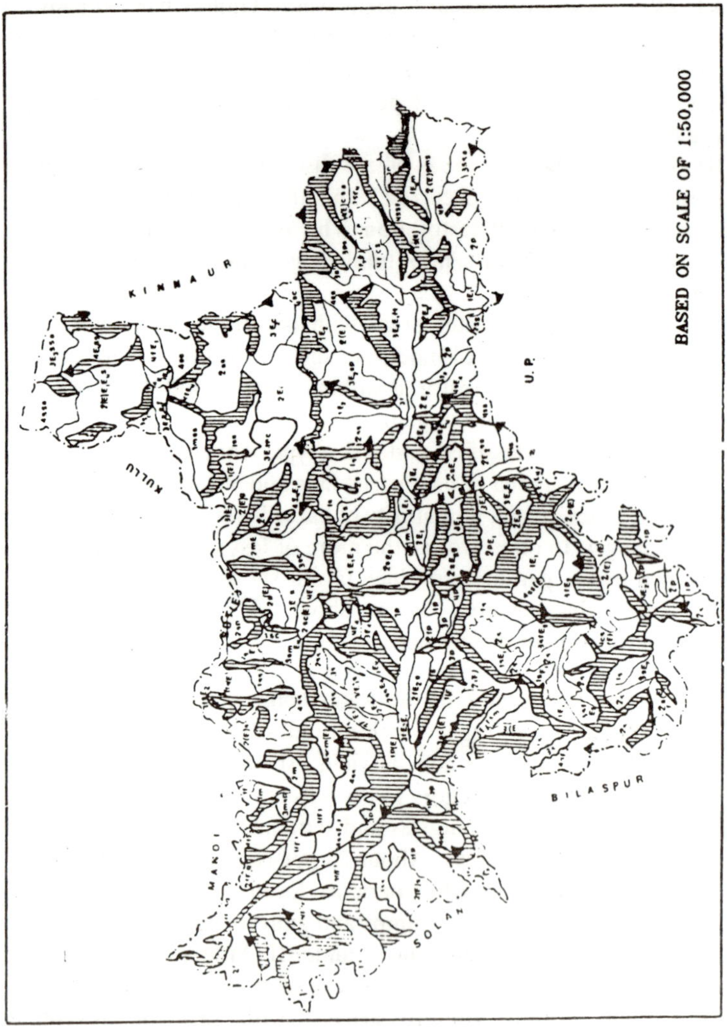

FIG. 5.6: Shimla District: Natural Hazard and Risk

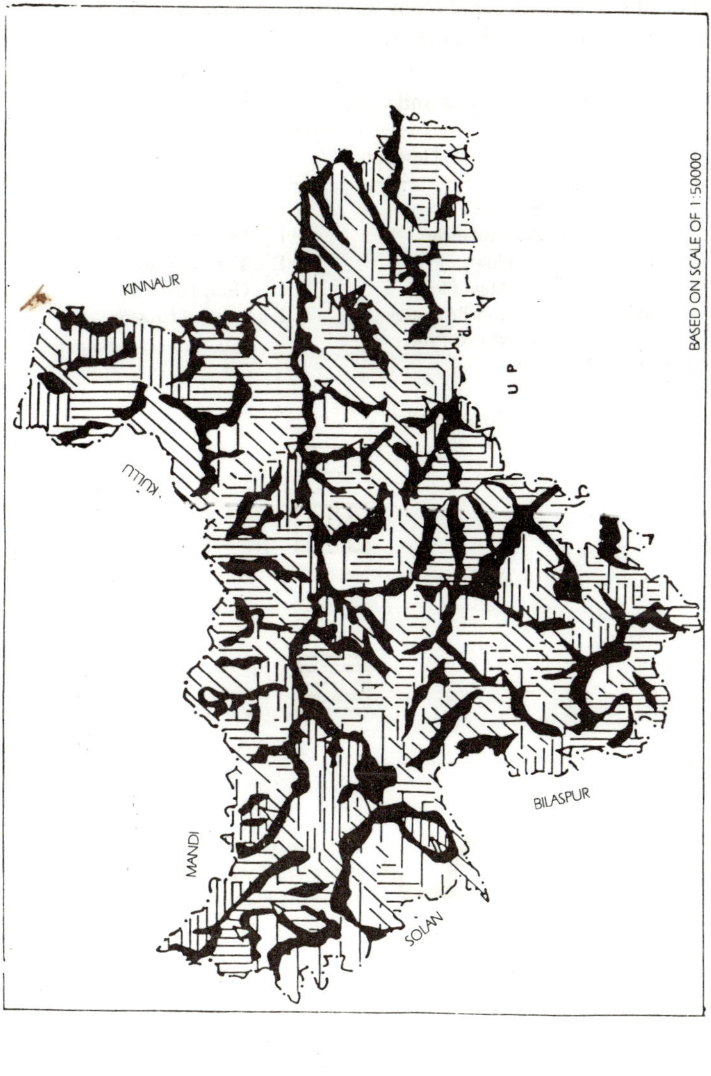

FIG. 5.7: Shimla District: Hazard and Risk Degree

An important method of representing information is to use symbols. Symbols are frequently used to show the type of event. These symbols seldom follow a true scale as they are meant to indicate a state rather than a physical entity. These symbols are pictorial representation of reality perceived from the study of hazards, the degrees of which were represented in Fig. 5.7.

Hazard Types and Consequent Risks

Symbol	Type	Description	
P	M,F,A	S : Slide	E_1 : Major Erosion
		A : Avalanche	E_2 : Lateral Erosion
SS	S	F : Flow	E_3 : Torrent Erosion
		M : Mass Flow	E_4 : Gully Erosion
E	F,M,A,S,E	P : Slow Processes	E : General Erosion
		SS : Rapid Processes	
E	E_1,E_2,E_3,E_4		

Risk Degree

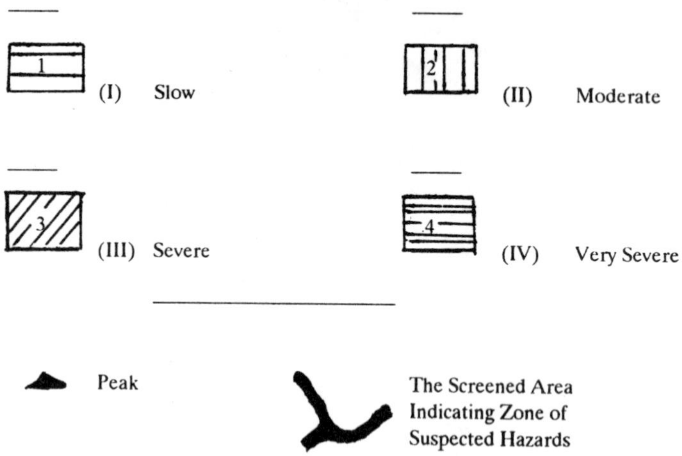

FIG. 5.8: Legend Index

As there is no accepted convention of hazard mapping symbols and as all hazard mappings tend to include an element of subjectivity, alphabetical initials' symbols approach was adopted to depict further details, which have been incorporated by using these initials in shades. A detailed information has been provided through description of the symbols in (Fig. 5.8).

Hazards in the zone are, of two types either *Active* or *Potential to Occur*. In potential to occur types of hazards, mostly tension cracks are the eminent signs of the unstable nature of the slope. In other words they forewarn the impending hazards. These cracks form direct paths for the deep infiltration, thus greatly minimizing the stability factor. If active, the symbols were not screened letting them represent the respective hazards, otherwise a shade was used. The mosaics of different active hazard types of the district, which depict the inherent instability of the area, are formed indicating that assessment of hazards in the area besides complexity is a highly difficult task. This aspect, due to differences in the hazard degree, which need individual assessment for accurate hazard avoidance and measures of suitability for various types of landuse, is made more intricate. With the help of this map, if a road alignment, passing through areas of different risk magnitudes is selected, thereby the hazard incidence risk can be assessed. Thus, identification of landuse and developmental activities would be permitted (Pirazizy, 1991).

Due to time constrain, while construction of hazard map, the past remnants of the hazards were compared with their present patterns. Meanwhile, some government establishments, such as Public Works Department (Shimla), Soil Conservation and Landuse Survey (Delhi), Forest Department (Shimla), and Technical Cadre of Snow and Avalanche Establishment (Manali), were consulted too. In addition to all above sources, the works of Kienholz et al, (1984), Fort (1987), and Ives and Messreli (1989) have been incorporated due to being highly applicable to the highlands in general.

5.8 **Analysis of Hazards**

The combination of, type and causal processes, is the basic condition for the classification of hazards and risks. The degree of risk is determined by the magnitude, frequency, and type of hazards. Human safety considerations were limited, and the main consideration was the occurrence and type of hazard because of natural topographic change or impacts on the morphology and terrain. Analysis of the hazard risk map stresses on water as the single major cause of instability in hazard areas. Thus, as a remedial measure, the map suggests, that if water removal or prevention of its ingress could be attained, then the hazards would have decreased in frequency, and number. In order to achieve such remedial measures, interception of surface runoff, minimizing of percolation, and establishment of moisture regimes in the areas with inclination are essential.

The areas of no hazard in the district, are confined to the slopes with vegetation cover, where obviously less distablizing processes, and therefore fewer instances of active hazards, within the large forest tracts of the district, can be observed. The hazard map shows that major hazards are confined to the settlement belt of Shimla-Sarahan, Shimla-Mandi, Shimla-Solan, Shimla-Bilaspur, and river banks with braided channels subjected to lateral erosion during monsoon, consequently causing the failure of roads, bridges, and terraces.

Hazard zonation map can readily be used, with almost great accuracy, to delineate areas according the magnitude of risk eventuality, with the objective of defining and controlling landuse. For example, perusal of this map can aid in location of new roads, settlements, hydal power projects, and variety of other uses, with safe and economic solutions (Pirazizy, 1991).

Some hazards being severe, are considered moderate because they occur away from anthropogenic interference. Hazards depicting the low risk are those which occur on gentle to moderate slopes, having limited disturbance and damage superficially. Human induced hazards, are seen in excavation works for construction activities. These constructions cause erratic runoffs making the area locally prone to mass wasting. Hazards caused by the anthropogenic activities, are mostly in the form of slides, arising due to inadequate stability consideration of slopes affected by such activities. Proper surveys, design, and evaluation of slopes can reduce the hazard risk eventuality to a large extent.

5.9 **Postlude**

Landscape of Shimla district is dynamic with rapid occurrence of natural and anthropogenic changes, such as construction of new land uses. Amongst natural processes in Shimla district, which coupled with anthropogenic interference acquire a hazardous shape with consequent risks involved, water erosion, earth flows, and rockfalls, either individually or in combination are the most important ones.

Some of the anthropogenic activities have had an impact on increasing instability of slopes, making them susceptible to potential degradation by runoff through floods, sheet erosion, slides, etc.. Yet considering the fact that terraces on the other hand tend to resist the degradation impact of running water, Ives and Messerli (1981) termed "Natural Hazards" instead of mountain hazards.

The substance generated through landslides or ground failure from northeast to east and west of Shimla district, is often untransported weathered rock consisting mostly of sand, silt, gravel, and at places boulders. The debris reflects the bedrock and, its hydraulic and mechanical properties are subjected to change with change in lithology. However, the mentioned substance is highly fragile in behaviour and loses its strength if disturbed. Rainfall being the main factor in initiating failure is highly effective in southwestern parts of the district, where texture of soils make them vulnerable to water erosion. Despite some fair extent of information on precipitation, its link with landslide activity is not adequately researched.

One of the major strategies for mitigation of natural hazards or their retrenchment is the micro zonation approach (Kuroiwa et al, 1978). By dividing the area of investigation into different sections of incongruous hazard occurrence, and using this information for location of anthropogenic activities, there is immense possibility of drastically reducing future human and material losses. An improved micro zonation method is a helpful instrument to coordinate the groups engrossed in multifarious activities. Thus, information on natural characteristics of the site is the best way to protect anthropogenic environment, as the hazards usually cannot be precluded. To ensure and respectively plan for security and stability of environment, as far as anthropogenic survival and ventures are concerned, adequate information is needed on geology, topography (relief, soil, slope), landuse pattern, and quantity and magnitude of risk eventualities. Thus environmental geography attempts to portray the solutions to the natural events capable of producing greatest damage, and the impact of time on the modification of impact.(Valdiya, 1987):

The above measures, though not capable of preventing natural hazard occurrence, may prove effective in reducing the magnitude, severity, and frequency of such hazards.

SECTION IV

EROSION SCENARIOS

Soil survey is not a simple process of mapping discrete parcels of the landscape. Each soil property changes more or less gradually in both vertical and horizontal directions. Soil survey enables rational allocation of land to different kinds of uses. Yet, the main aim of any soil survey ought to be arriving at a land classification. A general purpose soil survey is capable of producing a map which can be interpreted for many different purposes. The rationale in this section is that soil and landforms are very closely related. Thus, soil patterns and landscape elements often coincide and knowledge of one allows predictions to be made of the other. This relationship has been used as the basis for a number of integrated surveys, which are valuable adjuncts to problems of environmental management.

Having surveyed and described the above factor, the next step is to determine the properties of land that will be used in suitability assessment. These can be said to be limitations or requirements, which are usually expressed and comprehended in negative sense. To evaluate these limitations land qualities play an important role. Land determining the suitability boundaries.

The content of any evaluation, both in soil survey and suitability measurement will vary with the objectives and circumstances of survey. The present section comprising of two chapters six (erosion appraisal), and seven (modelling and suitability), is an attempt to clarify and link the foregoing discussion.

CHAPTER SIX

ENVIRONMENT OF EROSION

6.1 Prelude

From the early human history, attempts to improve the natural environment for a better life has been spontaneous. A major part of this endeavour had been the reclamation of soil for food production. Most of the time this manipulation of environment had not desired eventuality, hence, leading to soil degradation. This degradation was inexorable due to lack of integral understanding of natural processes, systems, and their intrinsic dynamism. Any competent interaction with environment based on homogeneity and sustainability must be based on the comprehension of natural systems through their systematic appraisal (Pirazizy, 1991).

Human beings are dependent on soils[1], the consequential perspective of this dependency is the equilibrium of certain soil properties. Yet there are several properties of soil which are effortlessly modified, such as structure, chemistry, and organic content. In other words, anthropogenic activities are staged on the soil material which are created and modified by natural process (Trudgil, 1987). Anthropogenic impact on soil involves the impeachment or modification of vegetation cover through cultivation and management practices.

6.2 Appraisal Demurs

In many commission memoirs it is common place to find an exclusively most preferred explanation to the problem under consideration. This explanation is qualified as mandatory, through statements of assumptions and sensitives to change in inputs. In the present study, however, while the eventual aftermath of decisions about resource use is evidently of interest, yet propositions of any determination of accomplishment does not fall in the array of the investigation. Furthermore, the quality of the data used appears to be inadequate for cursory appliance as a basis for alternatives. Preferably, main concerns are with forms of analysis, identification of preferences, and instructive prerequisites for planning in these catchments. The contention in present chapter is solely concerned with physical phenomena, by demonstrating the estimates of water and silt yield in different catchments under distinct landuses. For the sake of simplicity and demur of data acquisition for other uses, only agriculture and forestry are accredited as landuses. Though, the whole of catchment is considered, yet the areas above main water reservoirs have been favoured preferentially (Pirazizy, 1991).

Soil erosion assessment research has broad and varied problems. The reason being that process of soil erosion is intermittent. Therefore, observation of erosion is impossible, broadly speaking, hence, leaving the researcher to investigate the consequences. Consequences of soil erosion are *Erosion forms*, *Eroded soils* (Zachar, 1982), and removed substances from the soil. The second perplexity encountered is the infallible assessment of erosion influence. For accurate appraisal, the prerequisite is that the eroded material should not be subjected to a long distance and duration transportation, since the erosion process is active. Inconspicuousness of erosion and obliteration of its traces is another paramount research hurdle. Eventually, the fact that erosion is not an isolated phenomenon and transpires with other factors, confounds the investigation dispensations, hence hindering them as a prime challenge. However, soil erosion research is the trial of mutually inclusive components, such as intensity of erosion, qualitative and quantitative corollary of erosion, susceptibility to erosion, effectiveness of erosion repression measures, and distribution of erosion phenomena. In the present chapter the main emphasis shall be laid upon susceptibility to erosion and distribution of various degrees of potential erosion over the area. Research of soil erosion is difficult for several reasons. The particular reasons affecting the present work are as follows:

- Soil erosion is an intermittent process, which implies that erosion act is not observable and only consequences are investigated,
- The greater the time interval, the difficult interpretation which means the greater transportation taking place, the lowered is the accuracy of interpretation and assessment of overall in-

fluence,

- Inconspicuousness of erosion which indicates that traces are rapidly obliterated as the original conditions are not known, it is impossible to determine erosion-caused changes,
- Recognition of annual changes is also difficult;
- Erosion does not occur as an isolated phenomena.

Notwithstanding these difficulties, soil erosion phenomena are assessed by a wide range of methods. The choice of a method depends mainly on the purpose of research.

6.3 **Erodibility**

The process of slaking i.e. formation of an amorphous slurry due to soil disintegration, which forms a paste of the soil, is the primary phase of soil damage. This paste covering the surface resists the process of water percolation. In case of inclination of surface the gravity flexes the slurry downwards at the rudimentary stages of erosion. Rills are formed as the flux of slurry intensifies due to increased volume of water. Moreover, different soil types have different proportions of getting damaged (Winter, 1974). Hence, according to Turk and Turk (1988), *Erosion is the process where by soil particles are removed from the ecosystem . . .* , when the rate of soil production is lesser than the proportion of soil removal. However, various systems have different rates of soil loss. Figure 6.1, correlates the dissimilarities in soil loss and runoff in four different systems.

A cognition of the processes accountable for the transition of soil and its governing factors on slopes is fundamental in rationalizing type and distribution of soils within any landscape. However, variation in soil properties is one of the major influencing factors in soil erosion. The term "soil erodibility" was first used by Cook (1946), and is assessed by measuring soil loss under fix conditions and by identification of soil properties. Two equally powerful agents causing soil erosion are water and wind. In the zone wind does not play a significant role in erosion process, thus, the emphasis shall be placed on water as an active mechanism in erosion of soil. Effects of water are depicted first through surface wash, followed by the effects of topography.

SYSTEM	UNEXPLOITED FOREST	GRASS	MILLET	BARE SOIL
SOIL LOSS BY EROSION TONNES	NIL	NIL	78.0	146.2
WATER LOSS BY RUN-OFF (%) OF RAINFALL	0.4 %	1.9 %	26.0 %	50.4 %

SOURCE : MODIFIED FROM TURK & TURK,1984.

FIG. 6.1: Comparison of Water Loss and Soil Erosion from Four Different Systems

Local saturations starts the flow when rainfall intensity surpasses the infiltration capacity, thus, commencing the process of surface wash commence, which varies in relation to vegetative cover of the slopes.[2] Values of soil erosion enhance when vegetative foliage is impaired, be it either through natural or anthropogenic impact. Besides vegetation cover topography can perhaps, be placed as the second dominant factor controlling the soil erosion through application of gradient parameters, i.e. slope length, steepness, and configuration. Any factor intensifying the downward pulling effects of gravity and downslope surface wash, is an active contributing ingredient of the erosion potential.

6.3.1 *Rates of Erosion*

In general erosion is considered as hindering food production system threatening the sustenance of anthropogenic environment. Increasing rates of soil erosion in the zone, since the commencement of agricultural activities has provoked concern[3] among various scholars who believe that the surface of the zone in aggregate has been deprived of about 1/3rd of its thin mantle of original top soil. The statistical arguments of districtwise soil erosion analysis have been in publication for quite a long period. The analysis is supposed to be of a micro-erosion potential/rate investigation type dominated by both natural and anthropogenic impacts, the details of which is followed in proceeding chapters. Yet, a generalized interpretation, primarily due to lack in data and keeping above assumptions and applying a little arithmetic monopoly, helpful in prognostication, presents an alarming picture of the instability of the zone.[4] However, about 42 per cent of land is being lost every year in India, with a national level average of 20 t/ha. of topsoil depletion. The percentage of land lost annually in the zone in many ways is believed to be quite near to the national level. However, steep slopes or rough terrain with considerable relief has hindered the development of characteristic soil zones which further complicates the erosion appraisal (Butzer, 1976).

Soil of the zone by the inherent natural vulnerability are highly capable of erosion and immensely prone to runoff over most of their spread for instance, Shimla district is physiographically comprised of lesser Himalaya, river terraces, and in between them slopes of moderate to steep (15–20%). The southern slopes are scarped and dip to northern structural valleys. This landscape is congenial for high runoff silt yield, because source material of soil formation are argillaceous and loose. The Phyllites, Schists, and Slates in the geological structure of the belt yield moderately fine textured soils, which contribute relatively more towards silt yield and soil erosion. Moreover, about 53 per cent of Sutlej and Giri catchments in Shimla district which account for .9946 hectare, is under severe erosion rendered with very high silt yield index of above 1300, and the area under high silt yield index of about 1200–1300 in the same catchments accounts for 6368 hectare that is 34.30 per cent of the catchment area. The figures for upper Beas catchment, from Manali to Pandoh Dam in Kullu district, indicate 10236 hectare for very high, and 12904 hectare for high silt yield index, which means 28.22 and 35.58 per cent of catchment area for each category respectively. Elaborate specification of soil erosivity and susceptibility cognition is endeavoured in chapter ten.

6.4 Methodology of Appraisal

In the process of soil erosion measurement various steps were followed which are as follows:

6.4.1 *Topography*

Topography is one of the natural factors, which is important in indicating soil erosion. This factor has two important parameters (slope degree and length), which indicate the relief conditions and influence erosion. Transportation of soil particles once dislodged, is profusely influenced but the degree of slope and its respective length. Values for the slope degree and length factor were derived, with modification to match the zone, from the equation used by Wischmeier et al. (1958). The slope category and length maps are as good representative of the prevailing relief differences in the region (Table 6.1).

The *LS* factor is calculated from the,

$$L = \{(l/22.1)3.5\} \, x \, \{[S = [0.43 + 0.30s + 0.0043s^2/6.613]\},$$

L = Slope Length in Meters, S = Slope Gradient in Percentage

TABLE 6.1: Values of Slope Length and Degree Factors

Slope Degre (%)	<5	5-10	10-15	15-20	20-25	25-30	30<
(*S*) Factor%	0.45	1.17	2.21	3.57	4.56	7.28	9.81

Slope Length (*M*)	5-10	10-50	50-100	100-250	250-450	450-750
(*L*) Factor	0.41	1.7	2.48	3.44	4.45	5.32

Climatic factor *D* is an expression of precipitation and its values considering the rainfall isolines of the region. The lower values of chart correspond with the lower elevations and vice-versa (Table 6.2).

TABLE 6.2: Values of Climatic Factor

Rainfall (MM)	<800	800-1000	1000-1200	1200-1400	1400-1600	1600-1800	1800-2000	2000-2500	2500<
D Factor	0.29	0.35	0.57	0.76	0.85	0.88	0.94	1.03	1.12

The precipitation erosion factor *D* is a measure of erosive force and rainfall intensity in normal year. Table 6.3 depicts the precipitation conditions, values, forces and intensities of the zone.

TABLE 6.3: Precipitation Characteristics

Annual Rainfall (mm)	Duration (min)	Intensity (mm/min)	Erosive Rain (mm)	Annual Erosivity (Joul/M2)
<800	60	0.09	40	960
1000	55	0.13	50	1200
1200	50	0.18	60	1440
1400	40	0.24	70	1680
1600	30	0.27	80	1920
1800	25	0.28	90	2160
2000	20	0.03	100	2400
2500	15	0.33	125	3000
>2500	10	0.36	145	3360

Sahu (1990), states that as the intensity is not uniform throughout the storm period then;

$$i = (F/2) + (1/t),$$

where, i = maximum intensity, F = total rainfall of storm in mm, and t = duration of the storm in hour, is best suited to Indian conditions. Moreover, intensity — duration — return — period characteristics for any zone can be calculated from,

$$i = \{KTa/(t+b)n\},$$

Where, i = intensity (cm/hr), T = return period in years, t = duration in hours, and values of *a, b, n,* and *k*, have been given as $k = 3.71$, $a = 0.2176$, $b = 0.38$, and $n = 1.0382$ in average for Northwest Himalaya.

Soil factor *P* is an expression of erodibility. This factor was derived from the consideration of soil type humus content and proportion of clay (0.001 mm) (Table 6.4).

TABLE 6.4: Values of *P* Factor for Varying Soil Types

Soil Group	Type	Organic Matter (%)	(P) Factor
Alluvial	Loamysand and Sandyloam	1.0	1.50
Brown Hill	Sandyloam to Clayloam	2.6	1.20
Non-Calcic	Loamysand to Clayloam	1.5	1.45
Brown Podzolic	Sandyloam to Clayloam	3.3	1.33
Greybrown Podzolic	Clayloam to Siltloam	1.8	1.33
Planosolic	Sandyloam to Sandyclayloam & Clayloam	3.1	0.90
Humus and Iron Podzol	Sandyloam to Sandyclayloam & Clayloam	5.5	1.113
Alpine Humus Mountain Skeletal	Loamysand to Loam	6.2	1.28
Brown Forest	Sandy-Clayloam	3.3	0.93

6.4.2 *Petrology*

To estimate the value of *G* factor or petrological factor, distribution of various rock groups of the region according to resistance and permeability were studied. The geological factor *G* in the region has a value between 0.9 and 1.5. Comparing the *G* factor with rainfall zones the following values were obtained for *G* factor from $G = (p2/p)$,

p = Annual Precipitation in Milimeter (Table 6.5).

TABLE 6.5: Evaluation of G Factor in Rainfall Zones

Precipitation (mm)	Permeability of Rocks	G Factor
<800	Moderate	1.1-0.9
800-1000	Moderate	1.1-0.9
1000-1200	Slight	1.3-1.1
1200-1400	Slight	1.3-1.1
1400-1600	Slight	1.3-1.1
1600-1800	Slight	1.3-1.1
1800-2000	Slight	1.3-1.1
2000-2500	Low	1.5-1.3
>2500	Slight	1.3-1.1

6.4.3 *Vegetation*

The values of *O* factor or vegetation factor were obtained from the consideration of the percentage of managed vegetation to natural vegetation (Table 6.6).

TABLE 6.6: Vegetation Factor

Managed Vegetation	0	5	10	20	30	40	50	60	80	100
(%) Natural Forest	100	95	90	80	70	60	50	40	20	0
O Factor (%)	0.2	0.25	0.3	0.4	1	1.22	2	2.5	3.2	4

This evaluation is based on the assumption that erosion diminishes under a full natural vegetation cover (i.e. forest) to fourfold than when the forest cover is removed

6.5 **Erosion Evaluation**

Investigation regarding the soil resistance to erosion is a determinant factor in establishing the susceptibility status. Analysis of various erosion factors and conditions, like the effects of precipitation, topography, vegetation, structural conformation, soils and anthropogenic activities can assist in establishing the susceptibility to erosion of the area. Land evaluation indicates the extent of suitability of land for specific land utilization type, and can be used to predict input requirement for a profitable and environmentally harmonious landuse, considering the feasibility and land conditions.[5] The present section shall emphasis on expected impacts of soil erosion on environment which are deciding factors in relevant utilization types. Moreover, different landuses have categorical land requirements, besides various levels of labour, capital and technological knowledge.[6] At the commencement of the study the impact assessment is determined through the general information, as is the case in the present inquisition. Later on after establishment of a model or framework, the further impact assessment research is bound to take the stock of more available knowledge for micro level analysis.

The actual conditions prevalent in the area strongly suggest that vegetation cover whether natural or managed, and length of slopes are mutually inclusive factors in the process of land evaluation.[7] Processing of data, field observation, and modification of attempts by scientist such as Hensch (1970), Zachar (1982), Stehlik (1970), Zdrazil (1965), Wischmeir (1959) and Smith (1965), to suit the natural environmental conditions of the zone led to devise of charts and tables, which depict generalized condition and approximately precise values of factors. The stock of data from which the values of certain site variables, i.e. user data, were obtained from the information provided in previous chapters according to the modification of the procedures of the above mentioned scientists.

Most of the fundamental data originated from initial data on slopes and their length (Fig. 2.5), precipitation (Fig. 3.2), and topography (Fig. 2.4), geology (Fig. 2.1), Soils (Fig. 2.2), and vegetation (Figs. 3.1 and 3.2). P value was developed with modifications from measurements of Stehlik (1970), G and D values from Zdrazil's (1965) work, L and S values from extensive works of Wischmeier and Smith (1965), and O factor from the measurement values of Zachar (1982). Most of the empirical data was obtained from the field surveys of Byers (1965), and Byers (1986, 1987a, 1987b,) on detailed soil erosion process studies at more than thirty plots in the Himalaya. Factors P, G, D, S and L of the equation represent the susceptibility of land to erosion process.

6.5.1 *Potential Erosion Risk Map*

The potential erosion risk map was generated by overlying the p^2/P erosivity index, slope angle, drainage density and the erodibility index of highland temperate soils (Fig. 6.2). This maps and other combinations of the physical parameters, were verified by quantitative observations made over years by RVP Catchments Ministry of Agriculture. To compare the map and field observations sample of areas from each erosion classes were compared to the qualitative observations of different erosion levels on bare soils, agricultural and other disturbed conditions using an interaction matrix classification technique. The potential erosion risk map was chosen for further investigation had a classification accuracy of X.

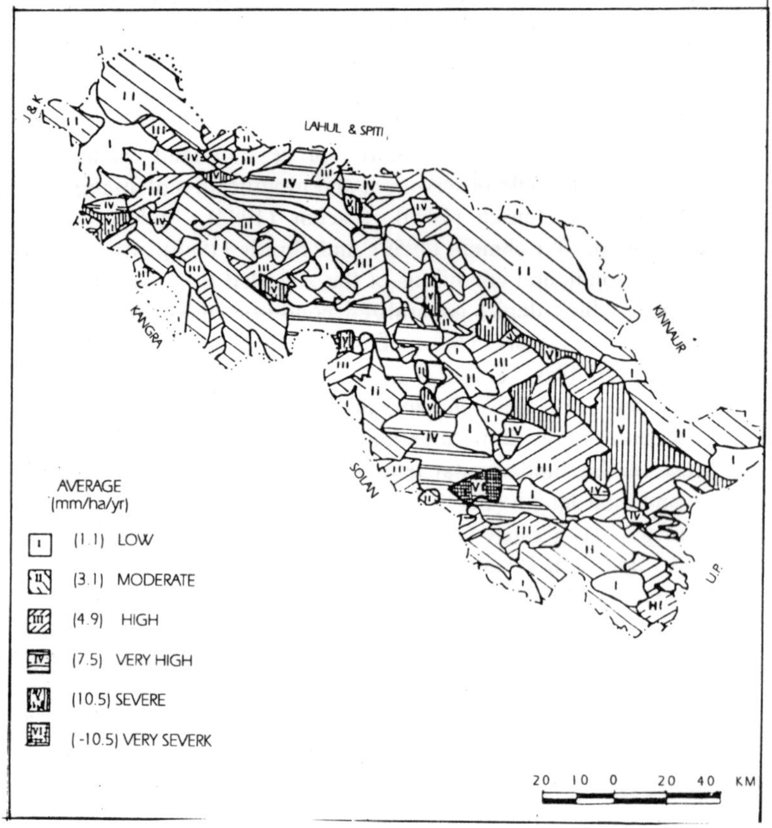

FIG. 6.2: Potential Erosion

$$X = \frac{\text{Units Sampled}}{\text{Total Units}} * 100$$

Till this point nature is the sole deciding factor of the extent *s* of susceptibility in various land units. Anthropogenic impact becomes evident through display and arrangement of vegetative cover in the form of various landuse types. Hence, factor *O* is the contrary to land susceptibility, rather it is determinant factor in increment or decrease in rates of erosion (Table 6.7) depicting the overall susceptibility of the zone to erosion and as it could be seen the range is quite vast i.e. from 0.5 mm/yr to 12 mm/yr (Pirazizy, 1991), various erosion potential categories in the zone are as follows :

Category (I) Low : The areas with low potential of erosion are confined mostly to gentler slopes (0-5%), of lower altitudes and more or less flatter regions. These are mostly situated in Chamba and Shimla districts. The areas under this category have a potentiality of about 1.1 mm/ha/yr of erosion (Fig. 6.3).

Category (II) Moderate : These areas, with an average 3.1 mm/ha erosion, are confined to slopes with moderate inclination (5-10%), and cover about 75 per cent of Chamba district (Table 6.7), Palampur and Kangra tehsils (Kangra district), about 43 per cent of Kullu district (mostly confined to glacial heights of northeast portion), and about 20 per cent of Shimla, and nine per cent of Mandi districts (Fig. 6.3).

Categories (III and IV) High and Very High : With annual potentiality of erosion ranging between 3.1–7.5 mm/ha/yr, covering about 65 per cent of wide spread through out the zone, situated on moderate to high slopes (10–25%), are the eminent centers of anthropogenic ventures ranging from production to heavy construction activities. These areas however, fall under naturally moderate to marginal suitable lands, as explained in the case of Fig. 6.2 in proceeding section (Fig. 6.4).

Categories (V and IV) Severe and Very Severe : With 7.5-10< mm/ha/yr, potential erosion, these are naturally highly capable of massive erosion and are confined to horticultural lands and overgrazed pastures, especially eastern Shimla and southern Kullu districts.

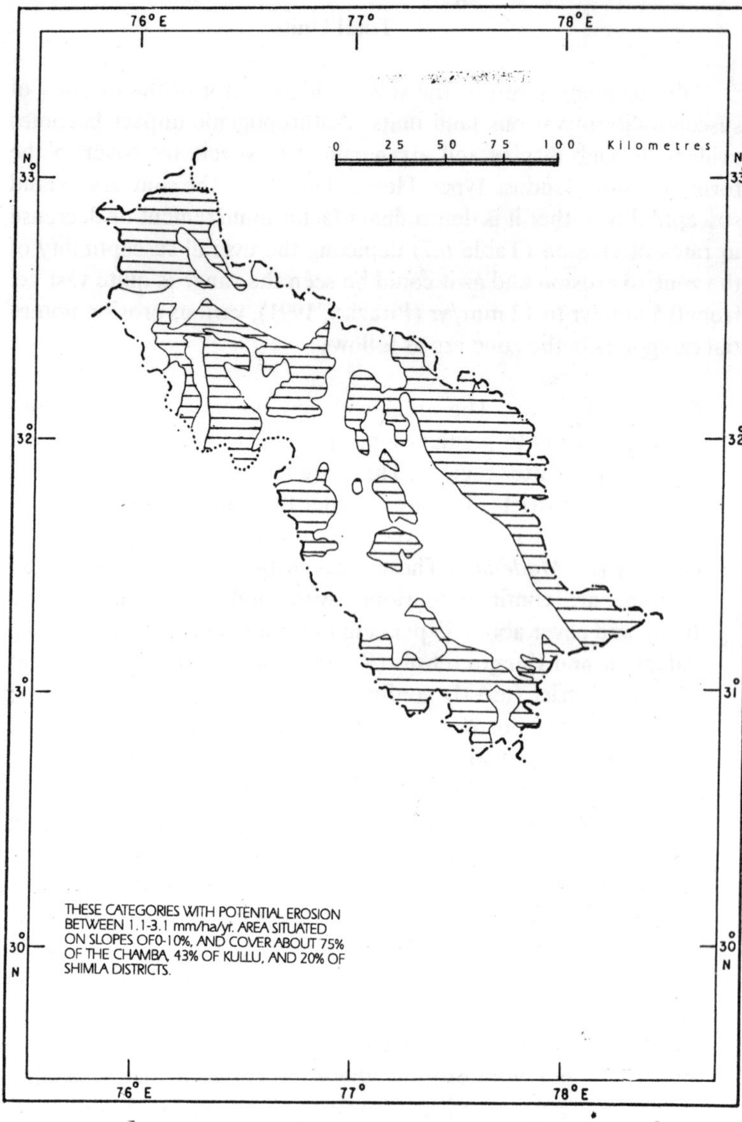

FIG. 6.3: Low and Moderate Erosion Category

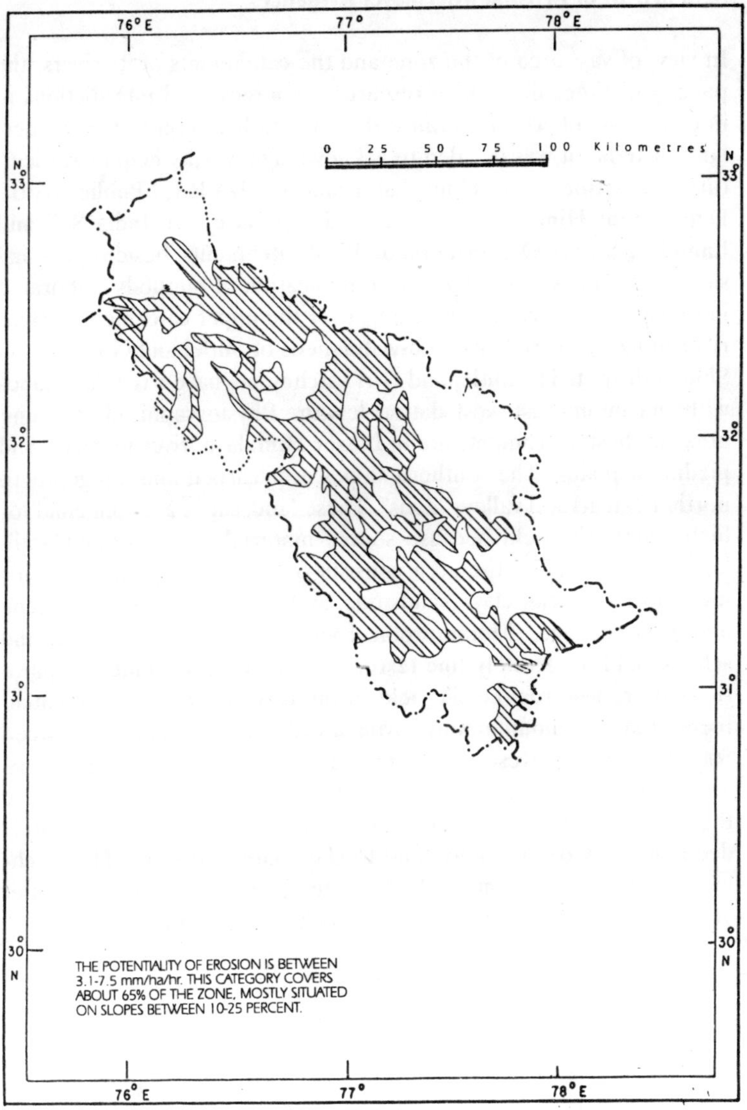

FIG. 6.4: High and Very High Erosion Category

6.6 Portrait of Erosion Microlevel Evidences

In view of vast area of the zone and the catchments of its rivers, the paucity of time, diminutive research resources, and interdictions in information collection, because the zone falls in Himalayan borderland of India, it was decided to acknowledge worthy help of Agriculture Departments of Himachal Pradesh, I.A.R.I, Public Works Department Himachal Pradesh, and reports of All India Soil and Land Use Survey Organization on RVP catchments, besides personal sampling. This section deals with application of methods elaborated in previous sub-sections preceding, over the upper catchment of Beas river in Kullu district, and sub-watersheds of Sutlej and Giri rivers in Shimla district. The Sutlej and Giri catchments have irregular shapes with maximum east west datum length. Physiographically the unit areas in these catchments are in Lesser Himalaya, river terraces, and piedmont plains. The southern slopes are scarped and dip gently to northern structural valleys "duns". These landscapes are congenial for high runoff silt yield because source material are loose and argillaceous. Geology of the area consists of alluvium, glacial deposits, metamorphic rocks, clays, sandstone, and conglomerates of Meocene and Pliocene age. Metamorphic rocks like phyllites, slates, and schists yield moderately fine textured soils and contribute comparatively more towards the silt yield in the catchment area. The catchment areas are mountainous having a wide range of altitude between 800-5500 meters. Reserved, demarcated, protected, private forests, agricultural waste, grazing lands, culturable lands, and orchards are the major landuses. In general soils vary from shallow to moderately deep and very deep, loamy sand to clay loam in texture. The catchment area is divided into 141 sub-watersheds out of which the unit areas considered comprise about 19000 hectares (Pirazizy, 1991).

6.6.1 *Spatial Characteristics*

The Beas catchment above Pandoh dam extends over an area of 4.99 lakh hectare, in an irregular rectangular shape. Physiographically it comprises of Lesser Himalaya, narrow valleys, and river terraces, embracing Dhaula Dhar and Pir Panjal ranges. Geologically the area consists of Colluvium, Glacial deposits, Quartzites, Shales, Sandstones, Granite gneiss, and Schists. The last two metamorphic group of rocks contribute to the coarse fractions in the soil cover.

The catchment is situated between 900-6000m. altitude, comprising of steep slopes, modified by agricultural terraces. Reserved, demarcated, and protected forests, agricultural wastes and grazing lands are the major landuses. In general soils vary from very shallow to moderately deep, sandyloam to silty clayloam and clay texture. The catchment area is officially divided into 230 sub-watersheds out of which the unit area considered comprise about 37000 hectares.

6.6.2 *Techniques of Organization*

For any volume of soil, or any unit of landscape, a researcher can make many measurements of soil properties (chemical and physical), and of landscape qualities (gradient, relief, and the like). The properties selected reflect the requirements of a specific purpose and the constraints imposed by impost, duration, toil, and access. Seldom any single measurement can be perused in isolation form those made by other specialists, and authenticate spatial anomalies of a given property as non-conformist of others. The procedure adopted, a modified outline of the scheme fostered in the *Bulletin of River Valley*, published by the Ministry of Agriculture (1972). To construct the base map, Survey of India toposheets, those not restricted, on the scale of 1:50,000 were used. The erosion intensity units were selected, implying a set of relevant parameters of soil erosion,[8] that exert combined reciprocal influence on soil detachment, as stated earlier. Primary reconnaissance of the units selected was done on site determinate locuses with the assistance of PWDHP specialized cadre.

Nevertheless, the sampling was conducted on sub-watersheds which are dominantly Lesser Himalayan landscapes in the upper confines and displaying Siwaliks in lower elevations. Depending on geology, slope, landuse management, soil, and erosion hazard peculiarity, these sub-watersheds were subjected to auxiliary divisions. All the drainage inquisitions are based on 1:50000 scale toposheets, ameliorating a systematic portrayal of multitudinous subwatersheds.

Empathetically, the delivery ratios were adjusted for each unit of scrutiny based on the data gathered from personal appraisal, factoral computations, and department of Agriculture IARI. This determinate ratio alludes to the percentage of the eroded material, which is carried either into the reservoir or debauchers into the flat valley plains (Fig. 6.5).

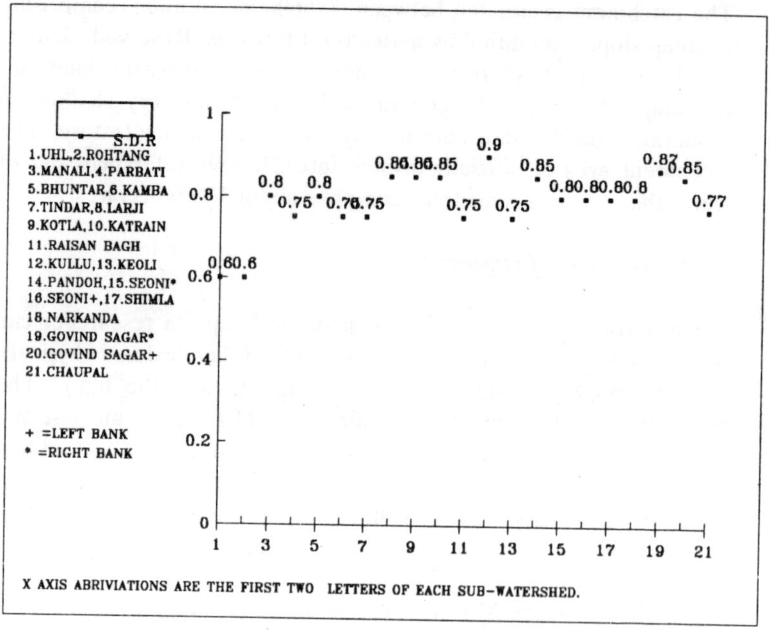

FIG. 6.5: Sediment Delivery Ratio Characteristics Major Sub-watersheds

To ascribe felicitous delivery ratios, shape and size of subwatersheds, relief, length ratio, physiography, slope, drainage characteristics, surface conditions, soil texture, and distance from active water bodies, i.e. lakes, ponds, and like, were premeditated. Ultimately, the area of each micro unit in subwatersheds was contrived by planimeter. The attained figure was multiplied by adjusted delivery ratio. The sum was divided by the total area of subwatershed and the product was multiplied by 100 to accomplish sediment yield index.[9]

6.6.3 *Stipulation of Inquiry*

The perimeters that make the erosion hazard prediction arduous, likewise promulgate economic evaluation of various resource management schemes strenuous. In addition, some catchment outputs such as conservation and anthropogenic ventures of exploitation, are not marked and cost of destruction inflicted are hard to value. Since, the objective of the study has been a critical probing into the potential erosion system analysis, the frailties in particular include the constraints of spatio-temporal interactions to felicitate linear programming approach, and the imperfections of the available data. The basic fault of the data is the inconsistency among certain appraises of silt and water yields from individual landuse activities. These limitations could, however, be quantitatively dilated to certain extent and not qualitatively removed, no matter how such a research is performed. Even, avoiding cerebral planning methods tends not to remove the limitations, but to substitute some of them for others that are not effortlessly specified with precision (Pirazizy, 1991).

6.6.4 *Erosion Hazard Peculiarity Units*

The higher values of silt load index suggest higher priority for sub-watershed conservation. Through the method of silt load index, four categories of erosion intensity have been selected to show the sub-watershed and erosion hazard peculiarity units. Catchmentwise analysis of erosion priority fixation for Beas, Sutlej and Giri rivers in districts of Kullu and Shimla are elaborated henceforth.

6.6.5 *Kullu District Catchment*

Generally the depth of soils in this catchment vary from shallow to moderately deep. They have a pale yellowish brown to dark brown colour, and sandy to silty clay loam and clay texture. Both calcareous and non-calcareous formations are prevalent. The soils are classified as Fsamments, Fluvents, Ochrapts, and Udalfs, of which moderately fine texture, shallow to moderately deep non-calcareous soils are most erosive. Soils of basin and valleys with medium texture and thick hydromoranic surface characteristics contribute comparatively less to the silt yield of the catchment. These soils are rich in organic matter and have well developed crumb structure that renders them less erosive both because of their characteristics in parts and partially by density of vegetation.

Out of 230 sub-watersheds in the upper catchment of Beas river, 14 sub-watersheds have been selected covering an area of about 36000 hectare. All the Upper Beas catchment covers an area of about 430000 hectare, area under snow cover is not contained. Considering the total selected units, about 28.22 per cent are under very high erosion intensity category with a silt load index of above 1300 tonnes (Table 6.7).

TABLE 6.7: Erosion Characteristics Upper Beas Catchment

Upper Beas Catchment	Area (ha)	Erosion Category Range	Silt Load Index tonnes	% Area to Total Catchment
UHL	1484	Low	<1000	4.09
Rotang	1496	Low	<1000	4.12
Manali	8628	High	1200-1300	23.79
Parbati	2604	Medium	1000-1200	7.18
Bhuntar	1712	Very High	>1300	4.72
Kamba	2112	Very High	>1300	5.82
Tindar	2048	Medium	1000-1200	5.65
Larji	1604	High	1200-1300	4.42
Kotla	1844	Very High	>1300	5.08
Katrain	2308	Very High	>1300	6.36
Raisan Bagh	2672	High	1200-1300	7.37
Kullu	3028	Medium	1000-1200	8.35
Keoli	2260	Very High	>1300	6.23
Pandoh	2468	Medium	1000-1200	6.80

High, medium, and low erosion intensity categories occupy 35.58, 27.98, and 8.22 per cent of the total number of sub-watersheds annotated. In aggregate, the upper Beas catchment has about 35.2 per cent of its sub-watershed units under very high erosion intensity category. Nevertheless, the figures for high, medium, and low are 33.9, 23.5, and 6.5 per cent. These very high erosion intensity category subwatersheds solicit a intendance on predilection rationale to instigate phenomenal decrement in the siltation (Table 6.8).

TABLE 6.8: Spatial Share of Erosive Intensity Category, The Upper Beas Catchment, Kullu district

Erosion Category	Silt Load Index Range	Area (ha)	% Area to Total Catchment	Number of Sub-watersheds
Very High	>1300	10236	28.22	5
High	1200-1300	12904	35.58	3
Medium	1000-1200	10148	27.98	4
Low	<1000	2980	8.22	2
Total		36268	100.00	14

As per conservation measures in the catchment, these subwatersheds require immediate cognizance, manifesting the tumult of predicament. Table 6.9, exhibits the proportionate distribution of erosion type units in each sub-watershed, providing an impression of the widespread characteristics of the catchment.

TABLE 6.9: Percentage Area Under Various Types of Erosion

Upper Beas Catchment	(vi)	(vii)	(viii)	(ix)	(x)	No Erosion
UHL	68.69	2.43	1.35	27.53	0.00	
Rotang	54.82	17.65	1.34	12.30	13.86	0.03
Manali	38.29	2.78	1.62	21.12	11.96	24.23
Parbati		12.44		87.56		0.00
Bhuntar			30.37	32.90	27.57	9.16
Kamba	10.15	2.65		75.13	12.07	0.00
Tindar	5.87	7.03	22.66	7.42	10.17	46.85
Larji	30.84	10.96	17.94	26.30	13.96	0.00
Kotla	16.71		21.69	33.62	27.98	0.00
Katrain	31.55		3.99	53.03	11.10	0.33
Raisan Bagh	25.91		13.47	37.73	22.89	0.00
Kullu	26.29	4.49	28.67	23.25	15.32	1.98
Keoli	12.57	16.46	19.29		14.87	36.81
Pandoh	37.60	19.13	14.26	6.32	12.16	10.53

6.6.6 *Shimla District Catchments*

The soils of this catchment in general have differing depths varying shallow to moderately deep in thickness. Their colour composition is dark brown and very dark gray. They have a loamy sand to clay loam texture, mixed with calcareous and non-calcareous components in contrasting proportions. Calcareous soils are only confined to Limestone areas. Soils of these sections are broadly classified as Fsamments, Fluvents, Orthents, Ochrepts, Udalfs and Udolls. Soils of river terraces, valleys, and peidmont plains, falling in this category, are less erosive and contribute comparatively lesser amount of silt yield to the river bodies. Soils under thick to medium forest cover are dark brown to very dark gray in colour and have a thick layer of litter and humus on the surface. These soils have developed granular and crumb structure in surface and subsoils. Infiltration and percolation rates are rapid, rendering these soils less erosive to some extent (Pirazizy, 1991).

Parts of Sutlej and Giri rivers draining the Shimla district, are sustained by multitude of tributaries situated in 141 sub-watersheds in an area of about 260000 hectare (Table 6.10).

TABLE 6.10: Erosion Characteristics

Sutlej and Giri Catchment	Area (ha)	Erosion Category Range	Silt Load Index tonnes	% Area to Total Catchment
Seoni RB.	1336	Very High	>1300	7.20
Seoni LB.	1886	Very High	>1300	10.16
Shimla	3944	High	1200-1300	21.25
Narkhanda	3924	Very High	>1300	21.14
Govind RB.	2800	Very High	>1300	15.08
Sagar LB.	2424	High	1200-1300	13.06
Chaupal	2250	Medium	1000-1200	12.12

From these sub-watersheds, seven major sub-watersheds have been selected, for reconnaissance, with an area of about 19000 hectare. Out of the total selected sub-watershed units about 53.58 per cent are under very high erosion intensity category. High and medium categories account for 34.30 and 12.12 per cent of the area respectively. Prevailingly, about 57.14 per cent of sub-watersheds in the appraisal units are of very high erosion intensity category (Table 6.11). Figures for high and medium categories are 28.57 and 14.28 per cent respectively.

As inferred earlier, the very high erosion intensity units in the district requisite an contiguous conservation-cum-treatment to intrinsically curtail already high silt load of the rivers.

TABLE 6.11: Spatial Share of Erosion Intensity Category

Erosion Category	Silt Load Index Range	Area (ha)	% Area to Total Catchment	Number of Sub-watersheds
Very High	>1300	9946	53.58	4
High	1200-1300	6368	34.30	2
Medium	1000-1200	2250	12.12	1
Total		18564	100.00	7

Table 6.12, elucidates the proportionate distribution of erosion types selected for probing, assisting in extensive characterization of the catchment and the operating landuses.

TABLE 6.12: Percentage Area Under Various Types of Erosion

Sutlej and Giri Catchments	(I)	(II)	(III)	(IV)	(V)	No Erosion
Seoni RB.	1.50	9.73	61.38	4.46		22.93
Seoni LB.	10.17	8.53	57.84	7.42		16.04
Shimla		5.17	26.27	63.18	5.38	0.00
Narkhanda	0.51	56.88	11.72	29.15	1.73	0.01
Govind RB.	8.57		91.42			0.01
Sagar LB.	0.83	99.17				0.00
Chaupal		16.70	44.40	38.90		0.00

6.7 Legend Description

If spatial variations occurred at random, without proper order, the holistic study as well as mapping efforts could be very difficult, due to the fact that information attained at one site could have little predictive value at new locations. In this case each site would be unique as there is no spatial order which could be cross-referenced from previous experiences in analogous settings. On the basis of a rapid traverse of the area and interpretation of toposheets, the mountainous landscape was sub-divided into two geological formation units, (i) Phyllites, (ii) Shales. These were further induced to ancillary divisions of incongruous erosion intensity units based on soil, gradient, vegetation, landuse, and erosional hazard peculiarity. These erosion intensity units are attributed roman numeric values, which are elaborated as follows:

(i) *Gentle Slope of Upper River Terraces, flood and piedmont plains, with approximately flatten slope (0-5%)*: These have exceedingly deep soil cover with medium fine texture at places affiliated with boulders and stones at exterior. The main landuse is medium to well exempted agricultural terraces, and slight to moderate erosion prevails,

(ii) *Gentle to Moderate Slopes (3–15%)*: These areas have a deep soil cover with medium texture. Agricultural terraces are well impregnable and erosion is slight,

(iii) *Moderate to Steep Slope (10–25%) order 1*: This category embraces moderately deep soil with medium to moderately fine texture. These areas are mostly cultivated with defective terraces, experiencing moderate to severe sheet erosion,

(iv) *Moderate to Steep Slopes (10–20%) Mountain Side or order 2:* At lower reaches, this category is constituted with moderate to shallow soil cover of medium texture on alternate Shale and Sandstone beds, having a thick forest cover and inefficient terrace cultivation. Erosion is slight in forest covered areas and moderate at terraced areas. At upper elevations slope is steep to intense (25–50%), with moderate soil cover. Texture is skeletal containing about five per cent bedrock outcrop. Grasslands are common landuse with overgrazing incidences at many stretches, and severe rill and sheet erosion is prevalent,

(v) *Steep Slope (>35%) Mountain Side:* This category comprises of remarkably steep, shallow to moderately deep soil of moderately coarse texture with 10–25 per cent bedrock outcrop. Moderate rill erosion and landslides prevail,

(vi) *Steep to Very Steep Slope (25–50%)*: Soil cover is shallow, with medium texture containing at places the bedrock outcrops. Dense forests are the only landuse. The area is covered with snow during winters. Erosion is slight to moderate,

(vii) *Very Steep Slope (>35%) order 1*: Soil cover is considerably shallow with gravelly medium texture consisting of bedrock outcrops. Major landuses are grasslands. The area is covered with snow during winters. Erosion operates moderately,

(viii) *Very Steep Slope (>30%) order 2*: Moderately deep soils with medium texture constitute this category. Inferior agricultural terraces are the only landuses. The area is covered with snow during winters. Moderate erosion is widespread,

(ix) *Extremely Steep Slope (>45%) order 1*: Very shallow soil cover with gravelly medium texture furthermore associated outcrops comprises this category. The area is covered with snow during winters. Grass cover is sparse. Quarries and landslides are chronic, and

(x) *Exceedingly Steep Slope (25-55%) order 2*: These areas contain very shallow soil cover, with medium textured gravelly associations with quite extensive bedrock outcrops. Moderately dense forests are major landuse. The area is covered with snow

snow during winters. Erosion is slight to moderate.

6.8 **Postlude**

A considerable amount of research and study has been undertaken on the mechanics of water erosion; this subject is now fairly well understood and documented. Similarly, a great amount of work has been devoted to developing different soil conservation practices and techniques. Unfortunately, the principles behind these subjects are not as widely known as they should be. The present indications with regards to population growth are that in near future large areas of new land will be brought under food production system. As most of the best land is already in use, considerable new area would be sloping with demurs of soil erosion. From long experiences it is known that particular soil is on particular slopes, in certain environments can only be safely farmed in certain ways. Once attempts are made to use these units of land in a way which exceeds their capabilities, inevitably a cycle of degradation and loss in productivity sets in.

This chapter has provided relevant information to determine general relationship between erosion and natural conditions, and their spatio-temporal variations. The final objective of this analysis has been an empirical estimation of extent and spatial dimensions of erosion, thus suggesting a systematic management to decrease erosion till a permissible limit within the natural suitability of environment, increase in soil fertility for sustenance of its vitality, and improvement in anthropogenic environment.

Soil erosion is a matter of concern to environment. Beside affecting food production and siltation of reservoirs, it is the basis of environmental degradation. Through soil erosion, the natural system is mobilized towards deterioration, setting the vicious hierarchy of chain effects into motion. The proper measures regarding soil conservation depend on comprehension of intricate relationship between land, its uses, its users, socio-economic and spatio-temporal environment. Though complex and at times prone to gross hypothesizing, these relationships are the basis of land evaluation. For a clear understanding of these relationships, the difference between land susceptibility to erosion and its degradation by erosion must be analyzed. However, the fact that land degradation, being a major land quality, is an efficient indicator in microlevel land evaluation, yet the land susceptibility to erosion is more relevant while considering mess-level land evaluation for application of anthropogenic activities. The environmental impact of soil erosion, till now has been interpreted as the impact of soil erosion on environment. This narrow definition has further been narrowed by mainly considering agricultural activities. Therefore, decisions on the feasibility of a specific landuse type, considering soil conservation, has been hindered by unidirectional approach. Soil erosion is an important parameters in regional planning. Soil erosion, moreover, directly influences the primary stage of land evaluation process, because in this stage the type of permissible landuses could be decided. Thus, it is concluded that, achievements of the potential of erosion hazard system analysis in catchment management, depends on an increasing awareness of its strengths and weakness and greater involvement by decision-makers in its appropriateness.

NOTES

1. Which have evolved from deposition of residuals at stages through the vigorous forces of their formation (Catt, 1979).
2. Vegetation through reduction of raindrops effect, prevention of compact layer formation, improvement of percolation rates, and impediment of surface erosion. It is observed that the humid temperate zones with complete vegetation cover and moderate rainfall intensities have extremely low surface wash rates (Young, 1960).
3. Different approaches to the problem is evident in the type of activity, its pattern, and shifts in patterns, which are circumstantial and vary in acute importance spatio-temporally.
4. I—One metric tonne of soil represents about 0.008 cm. of soil cover over an area of one hectare (Turk & Turk, 1988). II—Taking the annual erosion rate of 78 t/ha of crop (millet) land, and 146.2 t/ha of bare soil, the net annual soil loss amounts to 0.62 cm. and 1.17 cm/ha for both the categories respectively. As the rates of normal soil formation are very low and the work is of an approximate nature, the rates of Formation are not considered. III — For the reason that the zone is not uniformly distributed under various categories of figure (A), and at places the combination of categories is more expressive, the average erosion in the zone amounts to 0.90 cm/ha. IV – The average top soil of the zone is not anywhere exceeding the figure 18 cm. in depth. V – Thus, the total loss of the topsoil in the zone will be completed within the 20 years, if the processes are not accelerated in any way.
5. The land use types relevant to any region are consequent to socio-economic and general ecological conditions (Bennema, 1981).
6. While evaluating an area with regards to its land, there is an acute need for division of the area into land evaluation units, which have almost similar behaviour towards a specific landuse type. In the process of division of the areas into units, the land conditions which impart direct relation to requirements

ought to be comprehended. These conditions include, physical, ecological, social, economical and technological aspects. Land conditions in return determine the suitability of the land evaluation unit through the capability of land for application of various landuse types. However, a final check on the total impact of the land changes on the area is very vital (Vink, 1975).

7. Consideration of *L*, the slope length and *O*, the vegetation cover leads to the calculation of actual soil erosion which in turn assists the establishment of the land capability units. Therefore, the formula for actual soil erosion embeds *O* and *L* factors and is expressed as follows and is measured in mm/yr. $X = D\,G\,S\,P\,L\,O$.
8. These parameters are, physiography, slope(which controls amount and velocity of runoff), soil characteristics(which decide the potentiality of erosion), vegetative cover (which provides protection to soil), and land use (which indicates anthropogenic interference).
9. $$SYI = \frac{(Au \times DR)}{Asw} \times 100$$

SYI = Sediment Yield Index. Au = Area of Unit Considered
DR = Delivery Ratio.
Asw = Total Area of Sub-watershed.

CHAPTER SEVEN

MODELLING AND SUITABILITY APPROACH

7.1 Prelude

The basic procedure of conservation in the context of the zone (specifically speaking), is to carry out some form of land evaluation classification, which in return shall goad the planners to select an appropriate landuse system and, in necessary situations design conservation measures. The whole aim is to reduce the energy available for erosion which is mostly the function of two major factors i.e. water and gradient. Anthropogenic ventures are thus, directed at selecting landuses which result in less runoff while mechanical measures are adopted to hurdle the gradient of run.

The main purpose of the classification is to furnish basic informations on the use of the land along with general guidances on landuse and conservation practice. The classification of information on suitability map can be assessed qualitatively by identifying the most prominent areas of high erosion risk. This empirical approach, however, should be regarded as a short term expedient to be replaced eventually by a refined mapping system closely linked to deterministic models of erosion and conservation (Pirazizy, 1991).

7.2 Spatial Prognostic Evaluation

Any anthropogenic activity based on resource exploitation will be ambiguous in absence of erosion measurements and prediction for any area under consideration and cartographic representation pertaining to the degree of erosion. Himalaya, especially western Himalaya, are devoid of erosion maps, which make the classification of larger areas with regards to concepts and principles of susceptibility, extremely difficult. In the present section, based on data obtained from various sources and observations, generalized erosion maps are prepared to compare and correlate the intensity of natural erosion and its acceleration through anthropogenic activities. This section is solely devoted to natural erosion and presentation of maps and explanation regarding the mode they shall be utilized while discussing individual anthropogenic activity prevalent in the zone.

7.3 Model Scheme

Soil erosion is one of the most important of soil degradation hazards. Being a dynamic hazard, erosion assessment needs frequent updating. Reconnaissance-level erosion risk assessment in the zone of poses two pertinent scientific and practical problems.

- Eventhough a number of techniques are available as soil erosion rate and soil loss prediction tools, yet these mainly focus on the small scale. In the zone the requirement is for general assessments to be conducted over the entire zone as the soil resource base is often poorly known. Thus, problem lies in devising an applicable model which can be used to cope with prediction over large areas.
- Can this model account for different soil erosion processes active in the zone, focusing on actual demurs.

The development of a satisfactory quantitative measure of soil erodibility (susceptibility), is high priority, assessing the basic erodibility of soils in sites. Soils in this method will be quantified in terms of changes in the physical environment, which include climate, topography, geology, soil and vegetation cover.[1] Vegetation cover protects the soil from rain drops, runoff, dispersion, and increases the absorption capacity. Soil characteristics such as texture, humus content, structure and permeability are important in determining soil erodibility and susceptibility of soil particles to encounter soil erosion. A target figure then will be fixed. If the susceptibility (erodibility) exceeds the target figure, then through the model modifications would be prescribed to lower the soil loss. Though at times alterations would be major, but their contribution to soil loss problem can be anticipated. The above description vouches for the flexible and instructive nature of the model (Fig. 7.1).

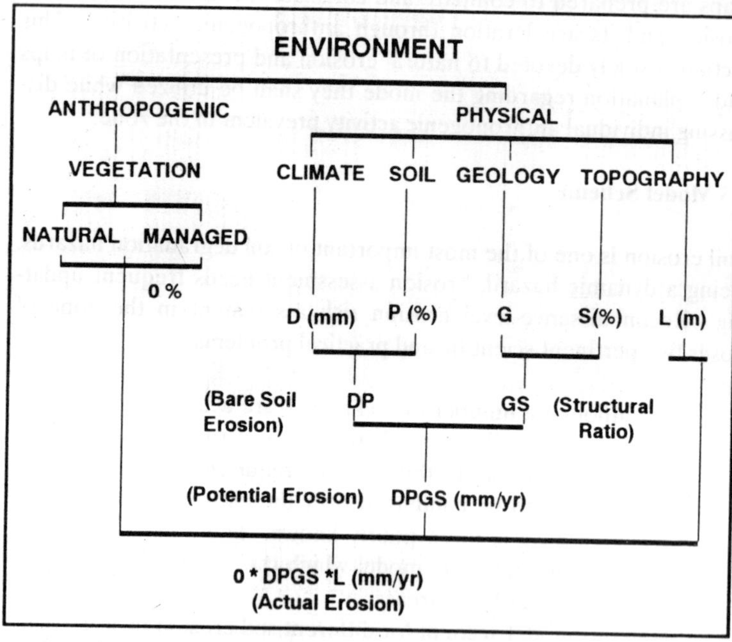

FIG. 7.1: Soil Loss Estimation Model of Temperate Zone Himachal Himalaya (Slemthh)

The modelling approach presents a comprehensive resource utilitarianism, based on specific purposes of limited resources, suitable to local conditions and harmonious to environment. The present model (SLETZHP), has been devised to provide an estimation of mean soil loss per annum. The model is considerably suitable approach specially in the case of highland Himalayan Temperate zone, where there is an inherent inability to assist costly research programmes, and urgent requirement for requisites of decision-making aid to encounter soil erosion. Figure 7.2, which is elaborated under Suitability Appraisal further in the same chapter, was prepared using the above model. This figure depicts various potential erosion categories prevailing in the Zone ranging from 5.1 mm/yr along the valleys and low lands to about 72 mm/yr on northwestern part, i.e. Pir Panjal Range in Chamba district and Dhaula Dhar Range. All the south facing slopes are the scenes of vigorous erosion. The model has been developed, besides facilitating the existing conditions of erosion and susceptibility of land to design procedures, based on empirical data and rational explanation, for protection and conservation of meager land available for production, resources, and encompassing environment. Due to paucity of time, sparsity of adequacy pertaining information, and eminent minuteness of the present (vital) problem, the selection of acceptable rational systems by combination of management practices is subjected to extensive specialized research.

7.4 Land Appraisal

The fundamental purpose of land evaluation is to predict the consequences of change. Prediction is needed of the land for different forms of use, inputs, management practices, and the consequences of such changes upon the environment. These include adverse consequences owing to a severe hazard of soil erosion. As stated by Dent and Young (1980):

> "Land evaluation involves comparison between the requirements of the land use and the qualities of the land. . . . Evaluation requires a comparison of benefits obtained with inputs needed."

Different kinds of landuse have different optimum conditions and different limitations. The clearest difference caused by the local conditions, which enhances susceptibility, is that of the potential of the land for use in specific ways, i.e. considering land characteristics which enhance land susceptibility to deterioration.

Susceptibility is the inherent natural factors of an area, which need field evidences. These factors are precipitation, soil, topography and geology. There have been various attempts to estimate and evaluate the susceptibility of the land to erosion.[2]

Based on these estimates various number of methods for assessing soil loss have been developed. These models are best suited for estimating erosion from very large areas. As the Zone is sufficiently a large area, and due to dearth of information and a permissible generalization, the modified equation of Frewert-Zdrazil method has been selected (Pirazizy, 1991).

The equation is as follows, $X = D\,G\,P\,S$, where X is the potential erosion (mm/yr), D the climatic factor, G the petrological factor, P the soil factor, and S the slope factor. This equation was invented to assess the rain wash erosion. In CSSR the method of Frewert (1955) was adopted by Zdrazil (1965) and modified by Stehlik (1970).

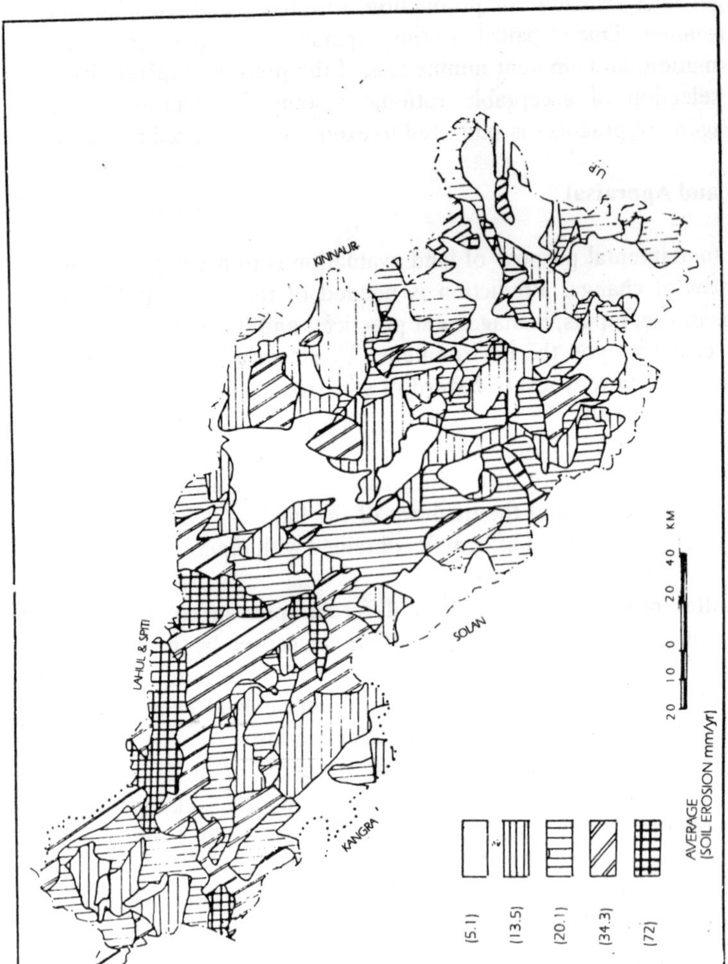

FIG. 7.2: Natural Suitability Classification

TABLE 7.1: Land Assessment Based on Suitability Kangra

	1	2	3	4	5	6	7	8
Rain Fall	2640	2640	2480	2480	3816	3816	1725	1600
Temp	18.2	18.2	20	20	17	17	14.5	6
Climate	WWM	WWM	WWM	WWM	WWM	WWM	MSA	MSA
Height	1219	1050	1170	2827	950	1250	1300	1800
Geo.	SS,GR	SS,GR	GW,SS	GW,SS	MU,CL	SH,SL	SH,SL	SH,SL
Landuse	CUL•	FST GLD	FST GLD	CUL••	FST	FST GLD	FST	FST GLD
Forest	MWT SAL	STR	STR	STR	MT SAL	MT SAL	MWT	MWT
Species	CKD BOFS	C	C	C	DFS BOBU	DFS BOBU	CKD BOFS	CKD BOFS
CL(%)	35.5	27.3	18.4	30.6	21.7	32.6	28.5	33.4
SI(%)	45.7	48.6	45.8	36.4	51.8	32.8	51.4	36.6

INDEX:
1 = PALAMPUR, 2 = HOTLA, 3 = PAROUR, 4 = TRIND, 5 = MELDEOGAP, 6 = DHARMSALA, 7 = KOTHI KORN, 8 = NAGNI. SI = silt, CL = clay, GR = granite, SS = sandstone, GW = greywacks, SH = shale, SL = SLATE, MU = MURRE, CUL• = cultivation (maize-wheat), CUL•• = cultivation (paddy-fallow), FST = forest, GLD = grassland, SAL = subalpine, M = moist, W = wet, T = temperate, STR = subtropical, C = chir, K = kail, D = deodar, FS = fir spruce, BO = ban oak, BU = birch, WWM = warm-wet mountainous, MSA = moist subalpine. TEMP = temperature (°C), GEO = geology, RAINFALL = (mm).

This modified method has been used for mapping large scale regional soil erosion. The equation of potential erosivity of the area is calculated considering the soil is unprotected by vegetative cover.[3] At the present conjunction through the development of potential erosivity map a capability map has been devised which is a relevant factor in deciding the suitable landuse types (Table 7.1).

7.5 Suitability Probing Scheme

The sequence of activities in conducting a suitability evaluation is shown in simplified schematic form in Fig. 7.3. The first need straddled on establishing the objectives of the evaluation. The available data was subjected to review, and assumptions, which were to be made, were formulated. This step was the planning for evaluation.

The field survey consisted of two parallel sets of activities, i.e. identification and description of landuse, and survey of available land resources. One of the fundamental steps in suitability evaluation was to consider various possible uses of the land as equal in importance as considering basic surveys of soil, landforms, precipitation, etc., having both aspects studied at the same time.

The kind of landuses were defined and described, which were set first by the objectives of the survey and second by the general physical nature of the area. The kinds of landuse were initially defined in very general terms. This was followed by determination of the landuse requirement and limitations. Simultaneously, the basic natural resource surveys were commenced, leading to production of the maps and other resource data on which the evaluation is based; these included a land systems or soil-land form map, precipitation data, vegetation types or other mapping units the relevant land qualities and characteristics to be used in assessing suitability were measured.

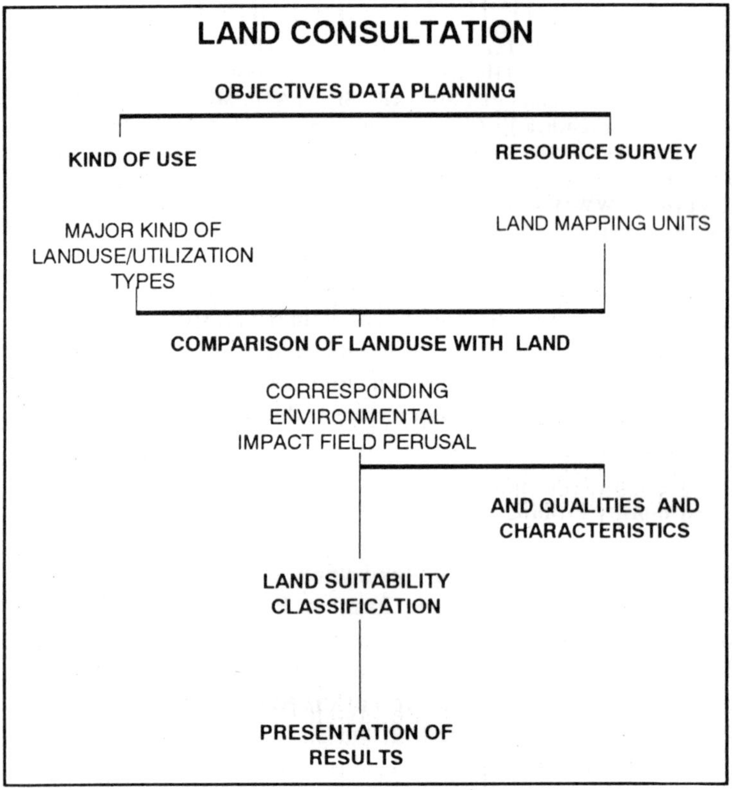

FIG. 7.3: Schematic Representation of Land Suitability Evaluation

The information collected by the resource surveys was that likely to be needed to assess suitability for the provisionally defined land utilization types. Thus, if it is clear that livestock or agricultural production is of importance, pasture or/and cultivable land resource surveys shall be included. The land suitability classification is then finalized, having proceeded by successive approximations through the later parts of the survey. This is then followed by presentation of results, i.e. map preparation. The essence of these procedures is the comparison between the qualities of the land and the requirements of the various kinds of use to which it might be suited. This is a major difference from the earlier approach to land evaluation, in which the resource surveys were completed first, and the resulting soil and other maps regarded as a fixed basis on which studies of land potential were built.

7.6 Land Suitability Appraisal

> "Land suitability evaluation is the process of assessing the suitability of land for specific kinds of use. These may be major kinds of landuse, . . .or land utilization types described in more detail,. . . . Suitability is assessed, classified and presented separately for each kind of use. . . . The results of a land evaluation survey are three-fold. . .[i.e.] "*Description* of land utilization types", . . . "Suitability maps", . . .[and] "Statement of the consequences favourable and unfavourable, of applying each kind of landuse to each area of land".
>
> (Dent and Young, 1980)

Amongst all three fold results of land evaluation, the second i.e. suitability maps are of eminent interest to the present book, specifically to the present chapter. These maps show that suitabilities of each land mapping unit for each defined kind of landuse. They may show on one hand that a particular mapped area of land is highly suitable for an anthropogenic venture, but not suitable for the other types, and on the other hand, these maps could depict that a particular mapped area is suitable for a specific anthropogenic venture, while the other mapped areas are not. For the sake of simplicity the present chapter contains a set of three maps and a tabular legend form presenting the information.

7.6.1 *Outline of Mapping Procedure*

While construction of suitability map (Fig. 7.2), he stress was that the evolution must involve a comparison of at least two kinds of landuses, thus, evaluation must be in terms relevant to the local condition and requires an assessment of the benefits to be obtained from a certain level of inputs (Gerrard, 1980). The classification is simple by qualifying the suitable category in terms of highly, moderately, marginally, etc. Thus, five levels of decreasing generalizations are defined. Land suitability classes reflect the degrees of suitability within orders. This scheme having been tested in Shimla district (Pirazizy, 1991), found to be quite successful. This is very similar to the specific interpretive soil map proposed by Wohletz (1968). These maps include agricultural land capability, soil limitations for various soil erosion hazards, production and recreational uses, and many more (Pirazizy, 1991).

Suitability classes indicate degrees of suitability within the order of suitability, the classes are defined on relative terms. The boundary, lower limit of highly suitable land, are set at the point where conditions have become clearly less than optimal, or inputs to counteract the effects of a limitation become needed. Suitability sub-classes indicate kinds of limitations. These limitations are indicated by lowercase letters placed after the class symbol, e.g. S_2s, S_3e, etc. There is no limit to the number of sub-class symbols that may be employed in particular survey.

7.6.2 *Rating Suitability*

Rating of suitability of land for agricultural activities in the zone is assessed by consideration of total biophysical factors, as discussed in chapter two and three, such as soil, slope, geology, relief, drainage, precipitation, vegetation, etc. Hence, application of the factors for assessment of fragility and suitability of zone can be seen from (Table 7.2.)

Human activities are nevertheless important in accelerating the natural processes and aid in swift modification of a landscape, subsequent to specific use patterns. In the zone under study, however, four broad categories of major utilization types can be identified which according to importance in shaping anthropogenic environment are as follows:

– *Settlement* : This landuse type includes construction and expansion of shelter, such as housing, hotels, markets and variety of other places.

TABLE 7.2: Ratingg Criteria of Suitability for Agriculture

Biophysical	Suitability Groups				
	(I)	(II)	(III)	(IV)	(V)
Topographical Zone	Flood Plains	Paddy Belts Dry	Forested Slopes	Alpine Pastures	Rocky Slopes
Geological Formation	Alluvial Deposits	Shale, Clay Limestone	Phylites Shale, Conglomerates Basic Rocks	Qurtzite Slates Graywack	Metamorphic Flysch Granite
Area (km^2)	1800	5800	7020	4000	1380
Altitude (m)	600-900	900-1800	1800-3000	3000-4000	4500<
Slope (%)	5-10	10-20	20-30	30-40	40<
Drainage	Poor	Moderate	Good	Very Good	Excellent
Soil Erosion (mm/ha/yr)	1-3	3-5	5-7.5	7.5-10.5	10.5<
Soil Depth	Very Deep	Deep	Moderate	Shallow	Skeletal
Soil Texture	Fine Loamy	Fine Sandy Loam, Clayey Silty	Coarse Clay Loam Clay Loam, Sandy Clay Loam	Sandy Loam Clay Loam Silty Loam	Bed Rock Exposed
(%) Topsoil Depletion	25	25-50	50-75	25-50	Nil
Vegetation	Sub-tropical	Oak, Broad-leaved	Coniferous Fir, Cedar Spruce	Pure Broadleaved	Glaciers Conifer
Rainfall (mm)	200<	160-200	140-160	100-140	80-100

– *Production* : This landuse type includes all activities which are related to production of food fuel and fodder, such as agriculture, horticulture, forestry, livestock, etc.
– *Construction*: This landuse type is the technocratic phase lan-

duse type which aims at enhancing accessibility, resource use, and expansion of transportation network, such as road, highway, and dam construction.

Preservation : This landuse type is exclusively upheld by the flourishing tourism industry which aims at increase of its resource base, i.e. aesthetic grandeur, natural vegetation, etc, to attain higher economic growth and beneficial returns by extending services (tourism infrastructure) to those interested.

Keeping the above factors in mind, an attempt was made to classify the landuse types (Table 7.3).

TABLE 7.3: Classification of Major Landuse Type with Regards to Suitability

Landuse	Suitability Groups				
	(I)	(II)	(III)	(IV)	(V)
Settlement	Good	Moderate	Marginal*	Serere	Very Severe
Construction	Good	Marginal*	Severe	Severe	Very Severe
Production	Good	Moderate	Marginal*	Severe	Very Severe
Preservation	Poor	Marginal*	Moderate	Good	Very Good

Note: Marginal* = It can be interpreted as either as not recommendable, or not suitable

7.7 **Predicted Suitability**

After the identification of erosion risk particularity units in chapter six, the potential erosion map was used as a basis for evaluation of suitability map. The potential erosion risk map shows the maximum soil loss. Thus, an adjustment was needed to account for these parameters which help in decreasing the potential soil loss, for example one of the very important parameters being vegetation cover. To proceed for evaluation of its map, three groups of variable were accounted for. These variables are as follows:

- Land use map with an scale of 1:50, 000,
- Population density/pressure map;
- Agricultural landuse map of those areas with high pressure having cultivation of maize, wheat, rice and barley as major crops

The best fit map was again assessed using observations indicated under potential erosion risk map section. Thus it can be said that erosion of soil is however, of two types either natural or accelerated. These two forms are of specific importance in erosion evaluation. Natural erosion is also referred to as geological erosion. The accelerated erosion is of two types, one naturally accelerated (this type is caused by draughts, avalanches, plant diseases, pest, etc.), and the other one is anthropologically accelerated. Thus, the main aim of the conservation and soil survey studies is to minimize, if not totally eliminate anthropogenic accelerated erosion to natural erosion.

7.8 **Suitability Appraisal**

One of the techniques of land evaluation used in soil conservation planning is the land suitability classification. The rationale of the classification system is that correct use of the land is the best means of erosion control. The system provides a method of assessing the extent and degree of limitations to the application of major use of the land, especially agricultural use, emphasizing particularly erosion risk. On the basis expounded and their severity the land is divided into suitability classes, each class representing a particular degree of severity of limitation. The suitability classification is must as most of the time the land is allocated to uses which were intended usually within a wider context of land appraisal and environmental management (Morgan, 1979). (Fig. 7.2), depicts that as the suitability is concerned the zone can broadly be divided into five suitability classes, which are as follows :

Class (I) Highly Suitable: With an annual erosion of about 1-3 mm/ha. these areas in general pose no significant frailty to sustained application of anthropogenic ventures. Settlement, construction and production landuse types are environmentally compatible landuse types in the areas falling under this group. Moreover, these areas are suitable for construction of infrastructure facilities in case of development of transportation network. These are situated in the valleys and flat plains with gentle slopes (5-10%). Prominent landuses practiced range from cultivation to construction activities, and agricultural terraces are well exempted. Soil cover is very deep with medium texture. This class covers about nine per cent of the total geographical area of the zone, mostly embracing Kullu-Manali valley (Kullu district), southwestern parts of Kangra tehsil (Kangra district), and southeastern parts of Shimla district, mostly along Sutlej and Giri rivers.

Class (II) Suitable: With an average soil erosion of 3-5 mm/ha/yr, these areas have frailties which in entirety impart little hindering to application of anthropogenic activities. Activities like settlement, construction and production are moderately compatible in environmental context. These areas are characterized with gentle to moderate slopes (10-20%), deep soils of medium texture, moderate erosion having a silt yield index of about 1000-1100. Major landuse consists of well impregnable and effective agricultural terraces. This areas approximately cover about 23.90 per cent of the total geographical area of the zone, and also can be termed as the minimum elevation areas, corresponding to the densely populated areas of foothills in Dharamsala and Palampur Tehsils in Kangra district, Chamba-Tissa tract (Chamba district), Kullu and Manali tehsils foothills and some tracts of Ani tehsil (Kullu district), Shimla, Rampur, and Chaupal tehsils (Shimla district), and Joginder Nagar and parts of Mandi tehsil (Mandi district) are places where this suitability class mostly prevails (Fig. 7.4).

Class (III) Moderately Suitable: With an erosion potential of about 5-7 mm/ha/yr, moderately steep slopes (20-30%), moderately deep soils of medium texture, moderate to severe erosion (silt yield index 1100-1200), and mostly forested at patches inefficient agricultural terraces, these areas bestow limitations which in aggregate are moderately severe for the application of anthropogenic ventures, like production and very severe for construction activities. The practice of the mentioned three landuse types in the areas under this category is not, however, feasibly suitable . The reason being that the areas have a critical environmental balance which can be disturbed by anthropogenic interference. Yet if some of the above land use types are practised, the preservation measures must dominate in order to mitigate the negative consequences. This belt also hosts major anthropogenic ventures with regards to major landuses such as, horticulture, terrace cultivation, and tourism industry, and also the most vulnerable belt which experiences majority of the land slides. This category is confined to the central portion of Siul Basin and Dalhousie in Chamba and Kangra districts between Dhaula Dhar and Pir Panjal Ranges, entire Kullu valley with exception of some patches near Manali, Bhuntar, and Larji in Kullu district, and Shimla, Narkanda, Jubbal, Rohru, Kufri, and Theog in Shimla district, Joginder Nagar and Barot tehsils and stretches of valleys in south of (Mandi district). This category covers about 35.10 per cent of the total geographical area of the zone (Fig. 7.4).

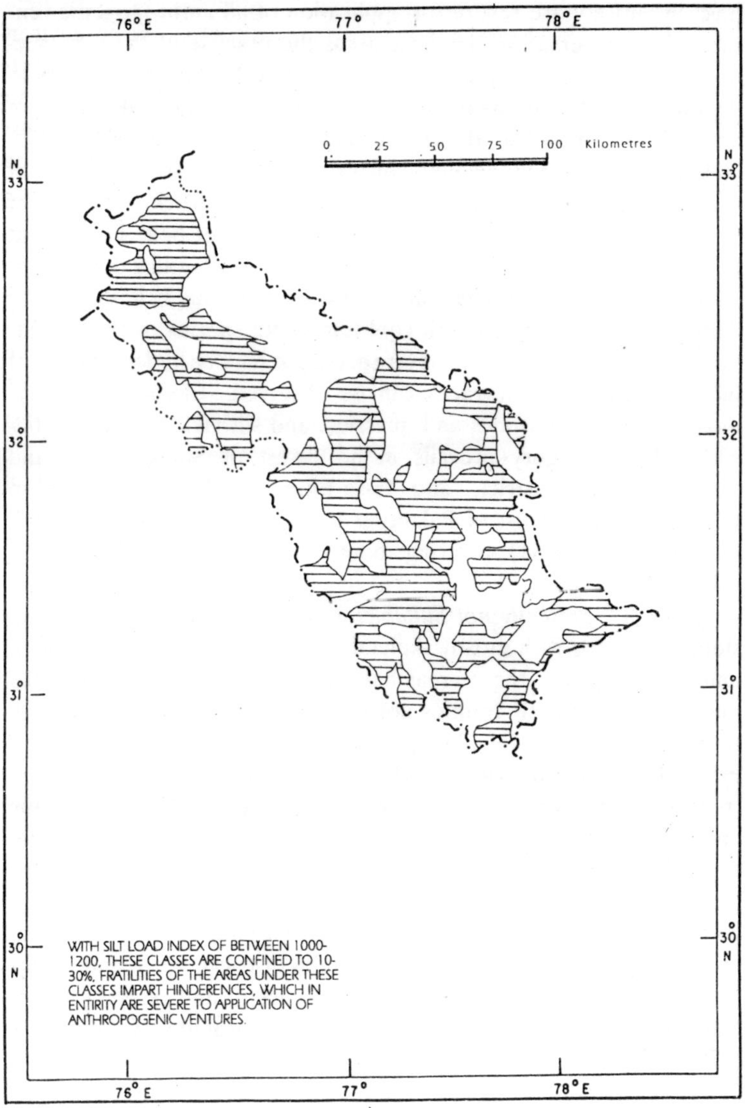

FIG. 7.4: II and III Suitability Classes

Class (IV) Marginally Suitable: The shortcomings that these areas pose in entirety are severe for application of all anthropogenic ventures, but preservation. In these areas the practise of the construction, settlement development, and production landuse types leads to degradation with varying intensity. With an average annual erosion of about 7.5-10.5 mm/ha., these lands are situated on steep slopes (30-40%), with generally shallow soil of medium texture and beds of shale and sandstone and about five per cent bedrock. Majority of forest area is situated in this belt, known as timber belt. This belt at higher elevations comprises the richest meadows utilized by Transhumants' herds for grazing, and forest cover in this belt changes to Alpine vegetation mixed with cool temperate species, with domination of *Betula* and *Rhododendron* species, at high altitudes patches of inferior agricultural terraces can be observed. Erosion is severe, mostly chronic landslides and quarries, and silt yield index of 1200-1300. This category prevails in northwest of Shimla, east and southeast Kullu district, and Tissa-Bharmour belt and southern Chamba district, most of Palampur tehsil (Kangra district), Mandi tehsil (Mandi district), covering about 29 per cent of total geographical area (Fig. 7.5).

Class (V) Presently not Suitable: Situated on steep to very steep slopes (40<%), with mostly skeletal shallow soil (10-25% bedrock), and coarse texture, these areas have an erosion order of above 10.5 mm/ha/yr, and their frailties are very severe that even individually they hamper any sustainable application of anthropogenic ventures, except preservation, which is of high value in tourism. Due to extremely steep slopes these areas are not at all suitable for any anthropogenic landuse activity. However, these limitations may be exceeded in time but currently not ostensible by the prevalent socio-economic and technical knowledge state. With an average areal coverage of about 6.9 per cent of total geographical area of the zone, the tracts of Chamba-Bharmaur (Chamba district), north and north-east of Palampur tehsil (Kangra district), and Shimla town, Narkhand-Rampur tract (Shimla district) are located under this class (Fig. 7.5).

7.9 Postlude

The basic procedure of conservation in the context of the zone is to carry out some form of land evaluation. The whole aim is to reduce

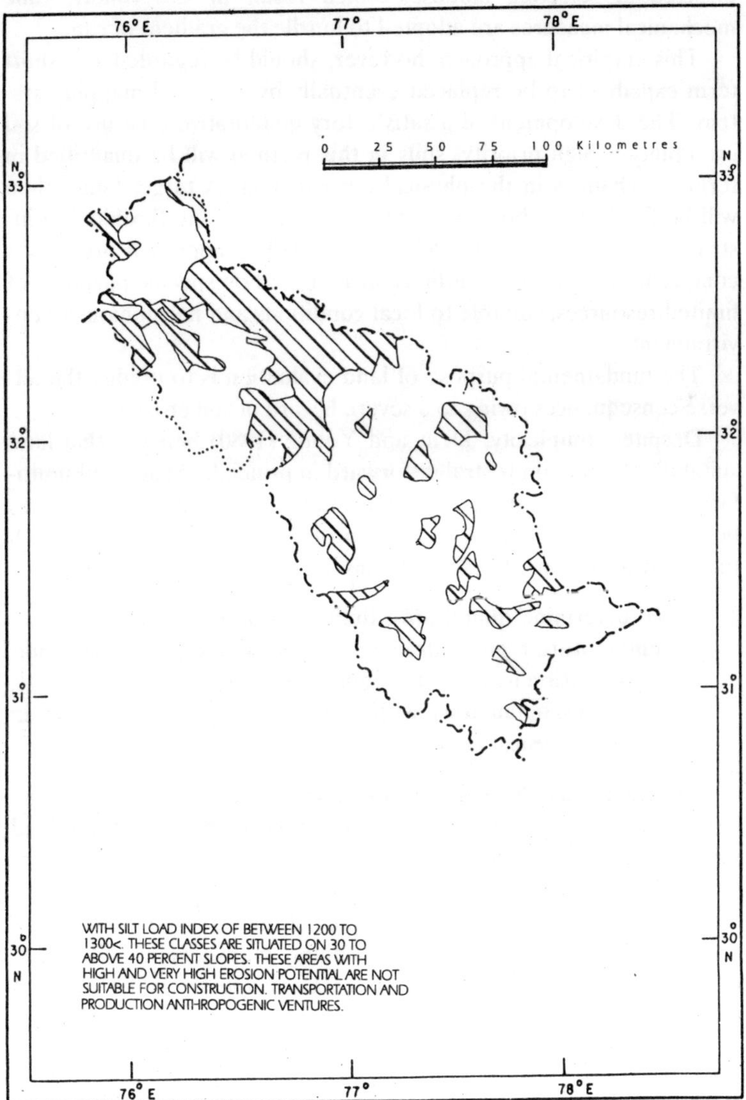

FIG. 7.5: IV and V Suitability Classes

the energy available for erosion Anthropogenic ventures are thus, directed at selecting landuses which result in less runoff while mechanical measures are adopted to hurdle the gradient of run.

This empirical approach, however, should be regarded as a short term expedient to be replaced eventually by a refined mapping system. The development of a satisfactory quantitative measure of soil erodibility is high priority, soils in this method will be quantified in terms of changes in the physical environment. A target figure then will be fixed. The above description vouches for the flexible and instructive nature of the model. The modelling approach presents a comprehensive resource utilitarianism, based on specific purposes of limited resources, suitable to local conditions and harmonious to environment.

The fundamental purpose of land evaluation is to predict the adverse consequences owing to a severe hazard of soil erosion.

Despite Complexity, Dent and Young (1980) believed that land suitability evaluation is straightforward in principle. Space and multifarious hinderences of research have not allowed all of the stages to be illustrated by examples. Yet, two problems require further indepth future investigations. These problems are :

- To determine quantitative limits for rating land qualities, i.e. temperature regime, drainage, nutrient supply, rooting conditions, erosion hazard, etc., applied to individual uses.
- To integrate analysis with suitability assessments made in terms of physical factors.

Though, failure or success of any land evaluation method is by the test of practical utility in landuse planning and management, yet, land suitability puts forward alternative possibilities for changes in landuse, whether major new uses or changes in detail, presents the consequences of such changes for each part of the surveyed area, and so provides evidence contributing to planning, management and investment decisions. Land suitability assessment appears to be of great applicability considering environmental conservation and sustainability development.

NOTES

1. Climate affects the potentiality of erosion both directly through the erosive force of rain, and indirectly through the parameters affecting vegetation

growth. Topography affects erosion through slope steepness and slope length which are critical factors in erosion potential due to their detrimental effect on the runoff velocity. Vegetation cover is the most effective measure to counter to soil erosion, hence, it is an important factor in calculation of soil erosion.

2. One of these attempts has been the universal soil-loss equation (USLE), developed by Wischmeier (1959), used in determining the total annual soil loss (E), with the following equation, (E = RKLSCP), where *R* is a rainfall erosion index, *K* is a soil erodibility index, and *L, S, C, P,* are factors of slope length, slope steepness, cropping practice and protection practice, respectively. The USLE is in practice over many countries to predict soil loss through sheet and rill erosion. Depending on the applicant, at times this equation is used for purposes not confirming to its basic objectives. It may be because the equation satisfies the need better than any other tool available. However, USLE being an empirical model, was developed by U.S. Department of Agriculture to estimate sheet and rill erosion from smaller areas i.e. agricultural land.

   In other words, the above equation determines the generalized susceptibility of the area to erosion. The units which have high susceptibility to erosion would be classified, as suitable, if a landuse type is protective to land, such as recreation resort and vice-versa like road construction.

SECTION V

PROGNOSTIC POSTLUDE

CHAPTER EIGHT

ABRIDGEMENT AND CONJECTURE

Environment is a dynamic open system, always in the process of change, irrespective of the place of prevalence. Long term changes in environment through geological evolution or/and short term modifications such as climatic fluctuations or/and the hazardous changes such as earthquakes, mass movement and the like, have always threatened anthropogenic survival and adaptation in highlands. The essence of the present investigation is not to focus on environmental problems. Rather it comprises of the consideration of the conditions that have been active in producing environmental demurs and constraints in their erasure/retrenchment.

Nature should be understood entirely through self consistency and not by reducing it into fundamental entities. In natural environment things exist by virtue of mutually consistent relationships, hence, all the studies pertaining to environment must be based on the fact that natural components of environment are consistent in inter, and intra-relationships. The most active environmental feedback in the technocratic phase of anthropogenic interaction has been morphogenetic. This feedback is in no way to lead to the original equilibrium or any equilibrium, rather it characterizes an environmental system with a certain relationships between altered natural constituents, either through natural or anthropogenic impacts or both, to be modified into another system with different relationships, characterized by lesser ability of its components to support biological life.

The present inquiry perimeter is comprised partially of the lesser Himalaya, constituted of platform type depositions, supposed to be the continuation of shield components, overthrust by southward nappes and thrust sheets, and partly of the higher Himalaya, depicting the line of great upheaval and maximum crustal attenuation and is the cardinal Metamorphic Belt. To evaluate the inherent natural vulnerability of structural conformation following empirical evidences support the conception of weak geological formation. For instance in Chamba district the Siul river sub-basin is comprised of Phyllitic Slates with interbedded Limestone and Dolomite/Limestone, and steep dip in northwest-southeast trend. These formations are situated in a major Kalhol Syncline which has weakened by two major faults along its core of fold axis, which makes the formation prone to discontinuity in structural planes, which result in slip occurrence during the shift in load factor. The example of weak geological structure in Mandi district is evident from Joginder Nagar to Pong Dam. The geological structure here comprises of weak structures of Sandstone, Claystone and Shales occurring in an alternate sequence. These bands are geologically considered as plastic bands which have low shear capability, thus leading to block and planar slides. The structural weakness observed in Shimla district is evident in lithological structure of Shalis, and Quaternary and Recent sediments of terraces complied with highly fractured and jointed rock compositions, which has rendered the Tata Pani-Dhar tract of Shimla district highly susceptible to landslides. The Dolomites and Limestones of this belt due to inherent weakness tend to develop depressions if the natural load factor is disturbed.

The soils of the zone vary according to micro-climatic conditions, perspective, and elevation range. These soils are alluvial to glacial in some tracts, and overall lacustrine in origin. The correspondence of soils to parent rocks is little, indicating a long geomorphic history, interrupted by alternative fluvial and glacial phases, as a responsible process in their evolution. The thickness of soil cover and its fertility status deteriorates with altitude. Due to steep slopes and comparatively low temperatures, which render soil forming processes slow, in higher altitudes rock beds are often devoid of soil cover. Natural vulnerability and susceptibility of soils to transformation has at present enhanced by anthropogenic interventions which perhaps, is conjecturing the supposed vicious deterioration syndrome, beyond the environmental resilience. Nonetheless, soils of the zone are classified with regards to various aspects of innate dynamism, from drought prone Loam to Clay Loams structure of medium to fine texture, mostly situated on the hilly slopes with great gradient variation, and deep fertile immensely acidic fine textured Silt Loam to Clay Loam, to the soils situated on higher elevations with meager agricultural suitability mostly covered with forests and grazing lands comprising of generally shallow Silty Loam to Loam texture and moderately acidic.

Soil of the zone by the inherent natural vulnerability are highly capable of erosion and immensely prone to runoff action over most of their spread. For instance, Shimla district is physiographically comprised of lesser Himalaya, river terraces, and over a major portion of its geographical spread slopes of moderate to steep (15–30%). The southern slopes are escarped and dip to northern structural valleys. This landscape is congenial for high runoff silt yield, because source material of soil formation are argillaceous and loose. Phyllites, Schists, and Slates in the geological structure of the belt yield moderately fine textured soils, which contribute relatively more towards silt yield and soil erosion. Moreover, about 53 per cent of Sutlej and Giri catchments in Shimla district which account for 9946 hectare, is under severe erosion rendered with very high silt yield index of above 1300, and the area under high silt yield index of about 1200–1300 in the same catchments accounts for 6368 hectare that is 34.30 per cent of the catchment area. The figures for upper Beas Catchment, from Manali to Pandoh Dam in Kullu district, indicate 10236 hectare for very high, and 12904 hectare for high silt yield index, which means 28.22 and 35.58 per cent of catchment area for each category respectively. An elaborate specification of soil erosivity and susceptibility cogitation has been endeavoured in chapter ten.

A cognition of the processes accountable for the transition of soil and its governing factors on slopes is fundamental in rationalizing type and distribution of soils within any landscape. Values of soil erosion enhance when vegetative foliage is impaired, be it either through natural or anthropogenic impact. Besides vegetation cover, topography can perhaps, be placed as the second dominant factor controlling the soil erosion through application of gradient parameters, i.e. slope length, steepness, and configuration. Any factor intensifying the downward pulling effects of gravity and downslope surface wash, is an active contributing ingredient of the erosion potential.

The relief and rock types of the zone are a part of its intricate and brittle abiotic environment giving it a variegated veneer, depicting a promiscuousness in the array of relief features. A variation in elevation from 950m. to almost 6000m. has hindered the formation of flat plains. However, relatively simpler geomorphic features are preponderant features of the altitudinal range of 1000-2000m., confining elaborate features of erosion land forms to 2000< meters. Elevations and inclinations are two important factors that differentiate between multifarious relief categories.

One of the major components of highland degradation with respect to anthropogenic utility impact is the steep slopes, which are under continuous modification by geomorphic processes thus, encumbering the formation and retention of soils. Agricultural terrace development on these slopes has reduced this inherent instability of slopes which range from gentle to extremely steep (35<%). The lower a slope angle, the deeper will be the soil cover with enhanced possibility of being covered by vegetation, and vice versa. Thus, any decrement in vegetative cover of the slope results in its amplified steepness and a shallower soil cover, which as a natural relationship of morphological system emphasis on the association of various components of system, and necessitating the complexity of the direct and indirect connections.

The congenital features which have given the zone a conspicuously explicit ecological character, are evident in its phytogeography, which is generally in the form of forest biomes dominant in intensely rainfed areas, coalesced with shrubs covering its floor. Moreover, spatial extent and conditions of precipitation decide composition, type, and proportion of vegetation at varying altitudes. The vegetation types of the zone in general are mixed forests of *Oak-Rhododendron*, *Pinus Wallichianu, Picea Smithiana, Abies pindrow, A. spectabilis* and *Cedrus deodara*. The Sub-Alpine zone consisting of species like *Betula utilis* and willows is also present in this region. Combination of *Betula utilis* with *Rhododendron companulatum* and *Juniper Spp.* reach the altitudes of about 4200m. Generally in the western part, or the settlement belts the Chir Pine belt has suffered most from anthropogenic activities. Higher up Oak forests have also been degraded by the similar impact, giving place to *Rhododendron* species which have no economic and fuel values. However, the vegetation in the zone is distinctly zonated. In eastern parts on the warm and dry aspects *Pinus roxbourgii* prevails while on southern dry slopes mixed Evergreen forests are dominant. Bushes are specialities of northern slopes.

Estimates of agglomerate productivity of vegetation and the net change in biomass has always been decided through the temperature effect. With the increase in altitude there is a decrease in average temperature, humidity, and rainfall. The highest maximum temperature in the zone is experienced in June, and the lowest minimum in January. Altitudinal analysis depicts that, between 900-3500m. altitude, the average monthly temperature of summer season declines from 28°C in lower valleys of Mandi and Shimla districts to 16°C in inner valleys of Beas and Ravi in Kullu and Chamba districts. The subtropical and temperate climatic blend could be observed all over the zone within the 900-1500m. altitudes. Above this belt cold temperate type of climate establishes between 2500-3500m. altitudes. Above 3500m. altitude Sub-Alpine climatic conditions are experienced. Durinal variation of temperature is greater during summer season especially on southern slopes. Length of cultivation and growing season is decided by temperature as is the limits for vegetation growth. These limits are favoured with precipitation distribution, enhancing anthropogenic occupancy and activities.

The relative humidity of the zone is decided through its active spatio-altitudinal factors, i.e. from outer to inner parts, southwest to northeast, average relative humidity decreases. Within the zone the precipitation increases with increase in altitude, yet various topographic features play a determining role in its microlevel variations. There is a spatial regression in precipitation distribution from outer south to inner north. The heaviest precipitation in the form of rainfall is received in Jogindernagar and Dharamshala strip of about 950 km^2. From Mandi the rainfall decreases with an average of 1250–2500mm. in the outer zone, 750-1250mm. in the middle zone, and 350–600mm. in inner zone around Pir Panjal range and beyond.

Considering temperature and rainfall distribution, the climatic pattern of the zone accommodates a pattern of wet sub-tropical belt including Dharamshala, Palampur, and Mandi, cool temperate belt at higher altitudes of Outer Himalaya including Chamba, Kullu, Dalhousie, Kharsog, and Shimla, and cold temperate and moist temperate belt of inner lesser Himalaya including Bharmour, Mandi, Rampur, and Kullu.

With exception of some portions of the Shimla district, which are drained by rivers Sutlej and Giri, the rest of the zone is drained by two major river systems Ravi and Beas and their tributaries. These two river systems with their tributary networks have carved two distinct catchments which are fed by precipitation, and have a northwesterly direction. Due to intermittent phase of crustal upheaval through recurring forces of orogeny, there is an immense variation in channel slants throughout the profile of both the rivers, thus, averaging the fall estimates of the total length seems as gross generalization. The normal decrement in gradient from head to base, is not a natural order in the Zone, as the river systems have to cut through several parallel ranges and are confined to the troughs in the intervals passing through ranges. These troughs decrease the velocity of rivers subsequent to plunge in gradient, hence, some depositional activities along their flow channels. During their flow through these troughs rivers have an average fall of 3 m/km. Both the rivers cut through most of the ranges, descending the next trough, which are located at differing heights, with a fall of about 19-38 m/km. For instance, the trough between the Great and lesser Himalayan ranges in the zone where Ravi flows, in Chamba district, is about 1524m. high, and the trough of Beas in Kullu district, is about 1250m., where as the trough between lesser Himalaya and Siwaliks for the Beas River in southern parts of Mandi district is about 850m. Moreover, the gradients of the River Basins have a marginal relation to their sources, because in Upper Himalayan system the average gradient does not escalate a great deal. Differences in physiographic and geological configuration of the zone has given rise to a diverse drainage pattern. Ravi catchment has an antecedent drainage of denderic pattern till Buntuh, from there till Mehla, streams acquire a semi parallel pattern. Drainage pattern of Siul River, a right bank tributary to the Ravi, is more or less rectangular till Choura where it joins the Ravi. Beas catchment on the other hand has a parallel pattern. However, after Dera Gopipur, the River starts a long lateral course in irregular lines, developing braided Channels, through constant shifting of Channels due to rapid deposition of non-cohesive material such as sand and gravel. Braiding, however, is a result of peculiarity of geometry, deposition, and sediment transport.

While discussing the anthropogenic impact, a distinction must be made between natural erosion, operating under the influence of normal geomorphological factors, and anthropogenically generated or accelerated erosion, a corollary of variegated anthropogenic ventures. Anthropohydrological interaction in the river system could broadly be grouped as direct channel manipulation, watershed modification, urbanization effects, and water pollution, of which the most conspicuous has been direct physical manipulation of river channels. Modification of watershed due to alteration of natural vegetation is another significant anthropogenic impact upon stream regions. Impact of anthropogenic activities on hydrological processes, which will be depicted in the performance of existent environment, is justified in massive volume of academic research. Many of these modifications such as increase in flood hazards, damages in river regime, causing atrophy in water quality, and reduced availability of groundwater, will instinctively actuate detrimental effects on antropogenic activities.

The causes of cataclysms vary from the inefficiency and inadequacy of drainage, their damaged conditions, ingress of rain water into the slopes through defective drains, and erosion by run off water. Among the major anthropogenic correlates in the zone are, improportionate urbanization, hasty and malexecuted transport network construction and expansion, and heavy location industrial activities disregarding the geolithological weakness prevalent over the highly elevated and variegated altitudes. In buildup areas, techniques of construction such as excavation, causes drastic modification of the slope, shifting stability to its critical frontier within a short while. Modern technology has rapidly concentrated large weights on the ground translocating surfacial material. All these modifications of slope induce conditions of instability, ultimately causing failure. As a result, what happens during a slide is reasonably adequately comprehended in strict physical terms. To probe into the combination of coercions invoking landslides, there is a requirement of the cognizance of both functional linkages between environment and slope stipulations, and ebullient of the environmental system. Among all the hazards associated to mountains, mass movement is the most serious one, which usually can not be prevented but can be predicted and their effects mitigated. However, the ultimate cause of all mass movements in the general pattern of landscape development. A landscape is a dynamic state, with remote chances of achieving equilibrium, mainly due to tectonic activities, upheaval of mountain massifs coupled with deepening channels of rivers, making all slopes comprising the relief to increase gradient, thus, resulting in mass movement as a disturbing factor in stability.

Today road designs stipulates a cut through steep slopes, eventhough, gentler slopes are preferable. Operation costs and land acquisition stress dictate construction on steep slopes especially at 25-30 degree. Seasonal saturation and weak vegetative cover, prone these slopes vulnerable to failure, and intense erosion. However, the road network in the zone exhibited an abrupt increase from 6000 to 6800 km. Only in two years period i.e. 1977-78 till 1979-80. Erroneous planning and obdurate execution of construction ventures has disestablized the generating massive debris quantity. The altitudinal dissimilarity of the origin and destination areas is another important component eliciting landslides. The main road systems in the zone originate and terminate at different elevations, with a diverse range of height of passage. The roads with low origin and high destination, vice versa, are the belts where large number of landslides occur. Despite the construction incompatibility with fragile ecosystems of mountains, most of roads have been subjected to widening to accommodate two way traffic.

Most of the landslides are concentrated between 1200-2500m. altitude, i.e. the altitudinal belt of 1200-1600m. (41.77%), 1600-2000m. (8.23%), and in 2000-2500m. (30.38%). The elevations comprising of weak geological structures of Shales, Limestone, Dolomite, Slates, Phyllites, and crushed Quartzites, are the stages of vigorous landslides.

Reservoirs, a modern venture in highland interactive systems, can be interpreted as anthropogenic counter force, which change unstable reverine ecosystems into stabler lacustrine ecosystems, arresting the discharge completely, which under less shear geolithological resistance of the zone can be disastrous. Moreover, stabilized water levels of reservoirs affect the energy flow regime, thus, reducing turbidity through depositional action, hence, intensified sedimentation.

Surface disturbances of physical environment are capable of producing increased erosion which in other words means greater concentration of sediment load in streams, the greatest contributor being construction ventures, thus weakening of the structure with consequences of landslides. Various deliberations depict that, the amount of suspended sediment concentration was recorded for streams passing through construction sites was 3000-150000 ml/l, while the corresponding figures for agricultural watershed was 2000 ml/l. Moreover, the mean figures for rural catchments are about 500-600 ml/l. This study concludes that, prior to the development of agriculture, mean figure was 50 ml/l, rising to 600 ml/l during intensive farming, and further increasing to 2000 ml/l during the period of construction activities. Landslides like other high energy processes, acquire hazard status, once anthropogenic environment is vulnerable. Main factors increasing the vulnerability of any area, in highlands, to landslide damages are urbanization, increasing reliance on transportation, and service systems. These factors are consequences of anthropogenic technical capability to modify the physical environment. In most of the instances, landscape anomaly has enhanced land crust sensitivity, increasing landslide frequency. However, in inherently unstable highland areas of the zone, frequent and extensive slides are common phenomena being quite lengthy and severe. The effect of debris produced through slides reaching the streams is mostly magnitude related, i.e., capability of sediment to be transported and its stability towards the action of fluvial activities.

Environmental risks in the zone, as an example, are almost all built area specific, with uncertain magnitude, frequency, and cause. The human choice factor to hazards is formed by the way hazards are defined. For instance, drought is a hazard as defined by a researcher, but rainfall variations are environmental dimensions according to a farmer in Temperate zone Himachal Pradesh. Thus, the individual or community response of farmers to spatio-temporal rainfall patterns are complex than a unidimensional set of adjustments to drought. A glimpse on the ethnography of the zone indicates that in the prevalent traditional culture of the zone, concern to classify hazards by their different effects is much less acute than the need to identify the chain of cause and effects, and till the genesis of effects are not found no adjustment is likely to be adopted by the locals. Moresoever, people's taxonomies are also different, for differentiating between anthropogenic and natural hazards, from the established research ones. Therefore, the spatial extent of hazards, appearing maladaptive, may be adaptive in the local context of its risks. Natural hazards, the major environmental risks, are however, the result of complex interaction between various natural systems, modify relief conditions as well as topographic gradients, through anthropogenic activities on a spatio-temporal scale. These local environments are made up of different sustaining components with specific resilience and inertia, well adapted to the local conditions. Thus, any change in physical configuration of any area directly affects its environmental sustainability.

Important characteristics in the geomorphic nature of highlands include significant structural influence in the landscape, through silt transportation in runoff, landslides, avalanches, rockfalls, glaciation, and talus creep. However, anthropogenic interferences alter the pace, frequency, and magnitude of these modifying processes. Yet, geophysical experiments have proven that the height of present time Himalaya is the same as what it was about a million years ago, with undiminished relief energy between crestline and plains. This proves the very insignificant contribution of modern human beings during the last few decades, when compared to natural balance created by dynamic exogenic and endogenic processes.

Natural erosion is the widespread phenomenon through out the zone, as an empirical evidence of natural hazards Shimla district was subjected to extensive hazard assessments. The geomorphology of the area displays two prominent features comprising of contrast between both slow and rapid (locally) processes, resulting in transfer of large quantities of substance. High slope gradient in topography is the basic causal factor, followed by geographic location, geological structures and composition.

Anthropogenic impact can be observed in the modification forms along the irrigated terraces, roads, settlements, pastures, leading to top soil loss on favourable substrata of shale, Litt, schists, and lacustrine deposits. These modifications of local relief from top to bottom of slopes is evident on many aspects, especially along the river valleys. Application of irrigation techniques, has directly affected water resource conditions. Thus, in general, most of the Shimla district can be considered as the anthropogenically modified landscape.

Landscape of Shimla district is dynamic with rapid occurrence of natural and anthropogenic changes, such as construction of new landuses. Some of the anthropogenic activities have had an impact on increasing instability of slopes, making them susceptible to potential degradation by runoff through floods, sheet erosion, slides, etc. The substance generated through landslides or ground failure is often untransported weathered rock consisting mostly of sand, silt, gravel, and at places boulders. The debris reflects the bedrock and, its hydraulic and mechanical properties are subjected to change with change in lithology. However, the mentioned substance is highly fragile in behaviour and loses its strength if disturbed. Rainfall being the main factor in initiating failure is highly effective in southwestern parts of the district, where texture of soils make them vulnerable to water erosion. The information on natural characteristics of the site and microzonation of hazards are the best ways to protect anthropogenic environment, as the hazards usually cannot be prevented.

From the early human history, attempts to improve the natural environment for a better life has been spontaneous. A major part of this endeavor had been the reclamation of soil for food production. Most of the time this manipulation of environment hadn't desired eventuality, hence, leading to soil degradation.

The environmental impact of soil erosion, till now has been interpreted as the impact of soil erosion on environment. This narrow definition has further been narrowed by mainly considering agricultural activities. Therefore, decisions on the feasibility of a specific landuse type considering soil conservation, has been hindered by unidirectional approach. Soil erosion an important parameter in regional planning, directly influences the primary stage of land evaluation process, because in this stage the type of permissible landuses could be decided.

Any anthropogenic activity based on resource exploitation will be ambiguous in absence of erosion measurements and prediction for any area under consideration, and cartographic representation pertaining to the degree of erosion. Western Himalaya are lacking adequate erosion maps, which make the classification of larger areas with regards to the concepts and principles of susceptibility, extremely difficult. It is widely acknowledged that, achievements of the potential of erosion hazard system analysis in catchment management, depends on an increasing awareness of its strengths and weakness and greater involvement by decision-makers in its appropriateness.

The paradigm giving the human beings an exceptional status is based on the fact that they are different from animals in their genetic and cultural conformation and the socio-cultural factors determine their interactions. Yet, these socio-cultural factors commence and are executed in the web of natural environments. Moreover, anthropogenic activities are staged on a finite biophysical environment with physical and biological restraint. The innovative nature of human beings and their attained power, according to the age old paradigm, can extent the carrying capacity limits of nature, yet natural laws of environment cannot be rescinded and anthropogenic activities can at least have unintended consequence.

Though the developmental activities create many environmental problems which are localized, yet the interaction between these problems gives them a regional dimension. Thus, the basic purpose of policies concerning environment and development must be concerned to enhance life quality along with satisfaction of human needs. In reality policies to promote anthropogenic ventures and designs to enact them are the forces directing impacts and consequences of anthropo-environmental interactive system, at a junction where far less is known about complexity of environmental systems prevailing in highlands. Besides, considering the natural rates of deterioration and its accelerated counterpart, assessing consequences of change and risks involved is quite intricate.

Environmental cause and effects are not linear systems, rather they are circular sequences which if approached unidisciplinary will generate paradoxes. Therefore, understanding the ways in which nutrients and other substances are transferred within the communities, and identification of storages of energy, and subsequents assist in, segregation of communities in biomass, focussing attention on large scale changes related to environment, and compatibility of anthropogenic utilization of environment.

The basic procedure of conservation in the context of the zone is to carry out some form of land evaluation. The whole aim is to reduce the energy available for erosion Anthropogenic ventures are thus, directed at selecting landuses which result in less runoff while mechanical measures are adopted to hurdle the gradient of run.

This empirical approach, however, should be regarded as a short term expedient to be replaced eventually by a refined mapping system. The development of a satisfactory quantitative measure of soil erodibility is high priority, Soils in this method will be quantified in terms of changes in the physical environment. A target figure then will be fixed. The above description vouches for the flexible and instructive nature of the model. The modelling approach presents a comprehensive resource utilitarianism, based on specific purposes of limited resources, suitable to local conditions and harmonious to environment.

The fundamental purpose of land evaluation is to predict the adverse consequences owing to a severe hazard of soil erosion.

Though, failure or success of any land evaluation method is by the test of practical utility in landuse planning and management, yet, land suitability puts forward alternative possibilities for changes in landuse, whether major new uses or changes in detail, presents the consequences of such changes for each part of the surveyed area, and so provides evidence contributing to planning, management and investment decisions. Land suitability assessment appears to be of great applicability considering environmental conservation and sustainability development.

SECTION VI

BIBLIOGRAPHY

CITATIONS

Agrawal, V.S. 1953 : *India as Known to Panini : A Study of the Cultural Material in the Ashtadhyay*, Lacknow.

Ai, N.S., 1985 : Effects of Neotectonic Stress Field on Landslides, *Proc. 4th. Int. Conf. and Field Worksh. on Landslides*, Tokyo, pp.433-436.

Aitkinson, E.T. 1886 : *The Himalayan Districts of the North Western Provinces*, Govt. Press Allahabad.

Ali, S.M. 1970 : *The Geography of Puranas*, Delhi.

Allen, J.L. 1972 : An Analysis of the Exploratory Process : The Lewis and Clark Expedition of 1804-1806, *Geographical Review*, 62, pp.13-39.

Anand, V.K. 1982 : A Note on the Structural Features of the Jutogh Formation in the Kullu Himalayas, Mandi and Kangra districts, Himachal Pradesh, *Geol. Surv. Ind. Misc. Pub.*, No. 41, pp.52-65.

Anderson, A. 1898 : *Final Report of the Revised Settlement of Kullu Proper*, Lahore Government Press.

Bajpai, S.C. 1970 : *The Northern Frontier of India*, Delhi.

Banon, H.M. 1952 : Fifty Years in Kullu Valley, *The Himalayan Journal*, XVII, pp.127-128.

Barsch, D. and Caine, N. 1984 : The Nature of Mountain Geomorphology, *Mountain Research and Development*, 4(4), pp.287-298.

Bateson, G. 1972 : *Steps to an Ecology of Mind*, Ballantine, New York.

Beckinsale, R.P. 1969a : River Regimes, In R.J. Chorley (ed.), *Water, Earth and Man*, Mathuen, London, pp.445-471.

Beget, J.E. 1983 : Radio Carbon-Dated Evidences of Worldwide Early Holocene Climatic Change, *Geology*, (11), pp.389-393.

Bennema, J. and de Meester, T. 1981 : The Role of Soil Erosion and Land Degradation in the Process of Land Evaluation, In R.P.C. Morgan, *Soil Conservation Problems and Prospects*, Wiley, Chichester, pp.76-85.

Bennett, R.J. and Chorley, R.J. 1978 : *Environmental Systems, Phylosophy, Analysis and Control*, Methuen and Co. London.

Bhardwaj, S.M. 1973 : *Hindu Places of Pilgrimage in India*, Thompson Press, Delhi.

Birot, P. 1969 : *The Cycle of Erosion in Different Climates*, B.T. Batsford, London.

Bolt, B.A. et al, 1975 : *Geological Hazards*, Springer-Verlag, New York.

Bowden, M.J. 1971 : The Great American Desert and the American Frontier 1800-1882 : Popular Images of the Plains, In T.K. Hareven (ed.), *Anonymous Americans, Explorations in Ninteenth Century Social History*, Prantice Hall, New Jersy.

Bowman, T. 1934 : *Geology in Relation to Social Sciences*, Scribner, New York.

Brandic, D. et al 1876 : *Suggestions Regarding the Demarcation and Management of the Forests in Kullu*, Calcutta, Govt. Press.

Briggs, D.J. and Smithson, P. 1985 : *Fundamentals of Physical Geography*, Hutchinson, London.

Bruce, C.G., 1914 : *Kullu and Lahoul*, London.

Brunton, P. 1937 : *A Hermit in the Himalayas*, Rider and Company, London.

Burmann, H.H. 1980 : *Shepherding in Dhaula Dhar*, Palampur, Indu-German Development Project.

Burrard, S.G. and Hyden, H.H. 1980 : *Geography and Geology of Himalayan Mountains and Tibet*, Part I, Gian Pub., Delhi.

Burrard, S.G. 1933 : *Big Games Hunting in the Himalayas and Tibet*, Gian Pub., New Delhi, 1980.

Burton, I.& Kates, R.W. 1964 : The Perception of Natural Hazards, in Resource Management, *Natural Resource Journal*, 3, pp.412-441.

Burton, I.R. et al 1978 : *The Environment as Hazard*, New York.

Butzer, K.W. 1976 : *Geomorphology from the Earth*, Hasper and Row, New York.

Byers, A. 1985 : Man-Induced Accelerated Surfacial Erosion, Upper Khumbu Valley Sagarmatha (Mt. Everest), National Park, Nepal, In T. Spencer (ed), *Abstracts of Papers for First International Conference on Geomorphology*, p.4.

Byers, A. 1986 : A Geomorphic Study of Man-Induced Soil Erosion in the Sagarmatha (Mt. Everest), National Park, Nepal, *Mountain Research and Development*, 6(1), pp.83-87.

Byers, A. 1987a : Landscape Changes and Man-accelerated Soil Loss : The Case of Sagarmatha (Mt. Everest), National Park, Khumbu, Nepal, *Mountain Research and Development*, 7(3), pp.209-216.

Byers, A. 1987b : An Assessment of Landscape Changes in the Khumbu Region Using Reapeat Topography, *Mountain Research and Development*, 7(1), pp.77-81

Carr, S. 1967 : The City of Mind, In W.R. Ewald (ed), *Environment and Man : The Next Fifty Years*, Indiana Univ. Press, Indiana, pp.197-231.

Catt, J.A. 1979 : Soils and Quaternary Geology in Britain, *Jour. Soil Sciences*, 30, pp.607-642.

Champion, H.G. and Seth, S.K. 1968 : *A Revised Survey of Forest Types of India*, Govt. Press Delhi.

Chapman, G.P. 1977 : *Human and Environmental Systems : A Geographer's Appraisal*, Academic Press, London.

Chew, G. 1985 : A Conversation with Fritijof, *Boothstrap Physics*, World Scientific, Singapore.

Chib, S.S, 1977 : *This Beautiful India Himachal Pradesh*, V(ed), Light and Life Publ., New Delhi.

Chib, S.S. 1977 : *This Beautiful India Himachal Pradesh*, Light and Life, New Delhi.

Chorley, R.J. 1969 : The Drainage Basin as the Fundamental Geomorphic Unit, In R.J. Chorley (ed.), *Water, Earth and Man*, Mathuen, London.

CLIMAP Project Members 1976 : The Surface of the Ice-Age Earth, *Science*, (191), pp.1131-1137.

Coates, D.R. 1985 : *Geology and Society*, Chapman and Hall, New York.

Coates, D.R. 1987 : Subsurface Impact, In Gregory, K.J. and Walling, D.E. (eds), *Human Activity and Environmental Processes*, Wiley, pp.271-304.

Cook, H.L. 1946 : The Infiltration Approach to their Calculation of Surface run-off, *Trans. Am. Geophys.*, Union 27, pp.726-43.

Cooke, R.U. and Doornkamp, J.C. 1990 : *Geomorphology in Environmental Management : A New Introduction*, Oxford, London.

Crozier, M.J. 1984 : Field Assessment of Slope Instability, In D. Brunsden and D.B. Prior (eds.), *Slope Instability*, Wiley, Chichester, pp.103-42.

Crozier, M.J. 1986 : *Landslides : Causes, Consequences and Environment*, Croom Helm, London.

Crozier, M.J. 1986 : *Landslides : Causes, Consequences and Environmental Perceptions*, Croom Helm, London.

Cunningham, A. 1924 : *Ancient Geography of India*, Culcutta, Govt. Press.

Dasman, R.F. 1976 : *Environmental Conservation*, Wiley, New York.

Deorr, A.H. 1990 : *Foundamentals of Physical Geography*, Wn.C. Brown, U.S.A.

Deterra, H. and Palerson, T.T. 1939 : *Studies on Ice Age in India*, Carnegie, Washington.

Dibble, R.D. et al 1984 : Volcanic Hazards of North Islands, Newzealand — An Overview, *Proc. Colloq. Earthquakes Geol. Hazards Predictions*, 27th Intern. Geol. Congr. Moscow, pp.144-59.

Directorate of Economics and Statistics 1988 : *Statistical Outline of Himachal Pradesh*, Government Press, Shimla.

Drew, Sir A. 1877 : *The Northern Barrier of India*, London.

Duthie, J.F. 1883 : *A List of Grasses of North-West India Indegenous and Cultivated*, Roorkee.

Eriksen, N.J. 1974 : *A Scenario Approach to Assessing Natural Hazards : The Case of Flood Hazards in Boulder*, Colorado, Univ. of Toronto.

Flew, A. 1986 : *David Hume : Philosopher of Moral Science*, Oxford Basil Blakwell, pp.176-83.

Ford, M.J. 1982 : *The Changing Climate, Responses of the Natural Flora and Fauna*, George Allen and Unwin, London.

Fort, M. 1987 : Geomorphic and Hazard mapping in the Dry, Continental Himalayas : 1 : 50000 maps of Mustang District, Nepal. *Mountain Research and Development*, 7(3), pp.222-238.

Fraser, J.C. 1972 : Regulated Discharge and the River Environment, In Oglesby, R.T. (ed.), *River Ecology and Man*, Academic Press, New York, pp.263-285.

Frewert, R.K. et al 1955 : *Soil and Water Conservation Engineering*, Wiley, New York.

Garner, H.F. 1974 : *The Origin of Landscpae : A Synthesis of Geomorphology*, Oxford Univ. Press, London.

German Fountation for International Development 1974 : *International Workshop on the Development of Mountain Environment*, Munich.

Gerrard, A.J.1981 : *Soil and Landforms : An Integration of Geomorphology and Pedology*, George Allen and Unwin, London.

Goldman, S.J. et al 1986 : *Erosion and Sediment Control Handbook*, MacGraw Hills, New York.

Golledge, R.G and Moore, G.T. 1973 : Symposium on Environmental Cognition : Introductory Comments, In W.F.E. Preiser (ed), *Environmental Design Research*, Vol.2, Hutchinson and Rose, U.S.A, pp.183-185.

Golledge, R.G 1967 : Conceptualizing the Market Decision Process, *Journal of Regional Science*, pp.239-258.

Goodall, D.W. 1970 : Studying the Effects of Environmental Factors on Ecosystems, In Reichle, D.E. (ed.), *Analysis of Temperate Forest Ecosystem*, Chapman and Hall, London, pp.210-222.

Gordon, W. 1914 : *Assessment Report of Harabagh Circle Mandi State*, Lahour, Govt. Press.

Gorrie, R.M. 1929 : A Short Description of Upper Bashahr Forest Division, *Indian Forester*, No. 55, pp.22-25.

Gorrie, R.M. 1929 : A Short Description of Upper Bashahr Forest Division, *Indian Forester*, 55, pp.534-539.

Gosh, D.K. 1980 : Geological Influence of Landslides and Other Mass Movements in the Planning of Baira-Baledh-Siul Link Project, HP, India, *Proc. Int. Symp. on Landslides*, Vol. I, pp.19-24.

Gould, P.R. 1963 : Man Against His Environment : A Game Theoretic Framework, *Annl. Assoc. Amr. Geog.*, 53, pp.290-297.

Grant, G.E. et al 1987 : An Approach to Evaluate Off-Site Effects of Timber Harvest Activities on Channel Morphology, *Proc. Symp. on the Effects of Forest Landuses of Erosion and Slope Stability*, Honolulu, Hawaii.

Gregory, K.J. and Walling, D.E. 1987 : *Human Activities and Environmental Process*, Wiley, New York.

Grene, M. 1963 : *A Portrait of Aristotle*, Faber and Faber, London.

Griffiths, J.F. 1968 : *Applied Climatology an Introduction*, Oxford Univ. Press, London.

Grove, J.M. 1979 : The Glacial History of the Holocene, *Progress in Physical Geography*, (3), pp.1-15, Edward Arnolds, London.

Gupta, H.K., 1976 : *Dams and Earthquakes*, Elsevier, Amsterdam.

Haigh, M.J. 1984 : Road Construction and Rural Stress in the Indian Himalaya, *Nordia*, 16(2), pp.175-180.

Hensch, B. 1970 : Water Erosion in Morocco, *Proc. Int. Water Erosion Symposium*, Vol. 3, Prague.

Holland, L.B. 1928 : *Report on Denudatyion and Erosion in the Low Hills of Punjab*, Lahore, Government Press.

Horrocks, N.K. 1961 : *Physical Geography and Climatology*, Longman, London.

Horton, R.E. 1932 : Drainage Basin Characteristics, *American Geophysics Univ. M.T.C.*, pp.350-361.

Horton, R.E. 1945 : Regional Development of Streams and Their Drainage Basins, Hydrophysical Approach to Quantitative Morphology, In Jervis, R.S. and Woldenberg, M.J. (eds), *River Network*, Hutchinson, New York, pp.80-99.

Hudson, N. 1971 : *Soil Conservation*, B.T. Bastford, London.

Hutchinson, J. and J.Ph. Vogel, 1933 : *History of the Punjab Hill States*, Govt. Press Lahore.

Hutchinson, J.N. 1968 : Mass Movement, In R.W. Fairbridge (ed), *Encyclopedia of Geomorphology*, Reinhold, New York, pp.688-95.

Hyden, H.H. 1904 : The Geology of Spiti, *Mem. Geol. Surv. Ind.*, 36(1), p. 129.

Ibbeston, D. 1883 : *Report on the Census of Punjab 1881*, Govt. Press Calcutta.

Ibbeston, M. 1914 : *A Glossary of the Tribes and Castes of the Punjab and*

NWFP Comp by H.A. Rose, Govt. Press Lahore.

ICIMOD, 1983 : Mountain Development : Challenges and Opportunities, Proc. 1st Intern. Symp. and Inaug. of the Intern. Centre For Integrated Mountain Development, December 1-5-1983, Kathmandu, Nepal.

Ittelson, W.H. and Cantril, H. 1960 : *Perception : A Transactional Approach*, Garden City, New York.

Ives, J.D. and Messerli, B. 1981 : Mountain Hazard Mapping in Nepal : Introduction to an Applied Mountain Research Project. *Mountain Research and Development*, 1(3-4), pp.223-230.

Ives, J.D. and Messerli, B. 1989 : *The Himalayan Dilemma : Reconciling Development and Conservation*, Routledge, New York.

Ives, J.D. and Messerli, B. 1989 : *The Himalayan Dilemma : Reconciling Development and Conservation*, UN Univ. Routledge, New York.

Ives, J.D. 1985 : The Mountain Malaise : Quest for an Integerated Development, In Singh, T.J. and Kaur, J.(eds), *Integerated Mountain Development*, Himalayan Books, New Delhi, pp.33-42.

Ives, J.D. et al 1987 : Research Strategy for the Himalayan Region : Conference Conclusions and Overview, In J.D. Ives and P. Ives (ed.) Proceedings of the Mohonk Mountain Conference—The Himalaya-Ganges Problem, *Mountain Research and Development*, 7(3), pp.332-344.

Jalote, P.M and Mithal, R.S. 1971 : Geological and Tectonic of Evaluation of the Recent Activity along the Krol Thrust in the Dun Valley. *Jour. Eng. Geol.*, 6(2), pp.421-428.

Jhina, P.S. 1962 : *Agriculture in the Hill Regions of Northern India*, New Delhi.

Jhingram, A.G., V.C. Thakur and S.K. Tandon 1976 : Structure and Tectonics of Himalaya, *Geol. Surv. Ind. Misc. Pub.*, 41 Himalayan Geology Seminar, New Delhi.

Johari, S.R. 1960 : *Our Borderlands*, London.

Joshi, K.L 1984 : *Geography of Himachal Pradesh*, National Book Trust, New Delhi.

Kates, R.W. 1963 : Perceptual Regions and Regional Perception in Flood Plain Management, *Papers and Proceedings of the Regional Science Association*, 11, pp.217-227.

Kates, R.W. 1970 : Human Perception of the Environment, *Int. Soc. Sci. Jour.* 22 pp.648-660.

Kaul, D.N. 1977 : Vegetation and Ecology of Indian Himalayas, In Acto die Colleque International, *Himalaya Ecologie-Ethnologie*, No. 268 edition, DN. CIVRS, Paris, p.155.

Kaur, J. 1985 : *Himalayan Pilgrimages and the New Tourism*, Himalayan Books, New Delhi.

Kayestha, S.L. 1964 : *The Himalayan Beas Basin : A Study of Habitat, Economy and Society*, Varanasi.

Kayestha, S.L 1971 : Himachal Region in India : A Regional Study, R.L. Singh (ed.), *A Regional Geography of India*, NGSI, Varanasi.

Kayestha, S.L. 1957-58 : Precipitation Characteristics of Himalayan Beas Basin, *Jour. Scie. Res.*, Vol. VIII No. 2, Banaras Hindu Univ. pp.183-89.

Khosla, G.D. 1956 : *Himalayan Circuit : The Study of a Journey in the Inner Himalayas*, London.

Kienholz, H. et al 1984 : Mapping of Mountain Hazards and Slope Stability, *Mountain Research and Development*, 4(3), pp.247-266.

Kirkby, A. 1973 : *Some Perceptions on Environmental* Hazard Research, Dept. of Psychology, Univ. of Bristol.

Knight, C.G. 1971 : Ethnography and Change, *Psychological Bulletin*, 70, pp. 47-51.

Krishnan, M.S. 1974 : Geology, In M.S. Mani (ed.), *Ecology and Biogeography in India*, Dr. W. Junk, B.V. Pub. The Hague, pp.115-132.

Krishnaswamy, V.S. 1980 : Geological Aspect of Landslides with Particular Reference to the Himalayan Region, *Roc. Int. Symp. on Landslides*, Vol. II, pp.171-185.

Kumar, Vanai, and Bhattacharyya, B.K. 1970 : Geological Mapping in Parts of Mandi District, H.P., *Geol. Surv. Ind. Misc. Pub.*, Vol. 24, No. 1, pp.35-48.

Kuroiwa, J. et al 1978 : *Microzonation Method and Techniques Used in Peru*, Proc. 2nd Int. Conf. on Microz. Vol. I, San Fransisco.

Kutzbach, J.E. 1981 : Monsoon Climate of the Early Holocene : Climate Experiment with Earth's Orbital Parameters from 9000 Years Ago, *Science*, (214), pp.59-61

Lagler, K.F. 1971 : Ecological Effects of Hydro-electrical Dams, In Berkwitz, D.A. and Squire, A.M. (eds), *Power Generation and Environmental Change*, MIT Press, Mass. pp.420-428.

Le Fort, P. 1975 : Himalaya : The Collided Range, Recent Knowledge of the Continent, *Amr. Jour. Sci.*, 275A, pp.1-44.

Low, B.L. 1968 : *Mountains and Rivers of India*, New Delhi.

Lowenthal, D. and Prince, H.C. 1964 : The English Landscape, *Geographical Review*, 54, pp.309-346.

Lowenthal, D. 1965 : English Landscape Tastes, *Geographical Review*, 55, pp.186-222.

Lowenthal, D. 1968 : The American Scene, *Geographical Review*, 58, pp.61-88.

Lyall, J.B. 1875 : *Reports of Land Revenue Settlement of the Kangra Punjab 1865-72*, Govt. Press Lahore.

Lynch, K. 1960 : *The Image of the City*, Cambridge, MIT Press, Mass.

Majumdar, D.N. 1961 : *Races and Cultures of India*, London.

Mani, M.S. 1978 : *Ecology and Biogeography in India*, Dr. W. Junk, B.V., The Hague.

Marwka, J.P. 1974 : Man's Impact on Stream Regimen and Quality, In I.R. Manner et al (ed), Perspectives on Environment, *Assc. Amr. Geog.*, (13), pp.80-99.

Medlicott, H.B. 1876 : Notes Upon the Sub-Himalayan Series in Jammu Hills, *Rec. Geol. Surv. Ind.*, Vol. 9, No. 2, pp.49-59.

Mitchell, B. 1979 : *Geography and Resource Analysis*, Longman, London.

Moore, G.T. and Golledge, R.G. 1976 : Symposium on Environmental Cognition : Introductory Comments, In W.F.E. Preiser (ed.), *Environmental Design Research*, Vol. 2, Hutchinson and Rose, U.S.A, pp.183-185.

Mukerji, A.B. 1986 : Landslides and Man, Development and Destruction in Western Himalaya, In Shafi, M. and Raza, M. (eds), *Spectrum of Modern Geography*, Concept, New Delhi, pp.397-428.

Narayana, D.V.V. 1987 : Downstream Impacts of Soil Conservation in the Himalayan Region, *Mountain Research and Development*, 7(3), pp.287-298.

Natural Hazard Research 1970 : *Suggestion for Comparative Field Observations on Natural Hazards*, Working Paper, Univ. Toronto, Toronto.

O'Riordan, T. 1981 : *Environmentalism*, 2nd. ed., London.

O'Riordan, T. 1986 : Coping with Environmental Hazards, In R.W. Kates and Burton, I. (eds), *Geography, Resources and Environment*, Vol. 3, Univ. of Chicago, pp.272-309.

Pande, I.C. 1975 : Recent Advances in Himalayan Geology, *Presidential Address I.S.C.A.*

Patnaik, H. 1981 : Role of Soil Conservation and Afforestation for Flood Moderation, *Ind. Natl. Sci. Acad. ND. Int. Conf. on Flood Disaster*, Mimeograph.

Peters, T.J. and Mool, P.K. 1983 : Geological and Petrographic Base studies for the Mountain Hazards Mapping Project in the Kathmandu-Kakani Area, Nepal, *Mountain Research and Development*, 3(3), pp.221-226.

Petley, P.J. 1984 : Ground Investigation, Sampling and Testing for Studies of Slope Instability, In D. Brunsden and D.B. Prior (eds), *Slope Instability*, Wiley, Chichester, pp.67-101.

Piaget, J. 1967 : Studies in Cognative Growth, *Contemporary Psychology*, 12, pp.532-533.

Pirazizy, A.A. 1989 : Impact of Construction Activities on Environmental Changes in Kashmir Himalayas, In I. Mohan, *Environmental Issues and Programmes*, New World Environment Series, Ashish, New Delhi.

Pirazizy, A.A. 1991 : Assessment and Cartographic Presentation of Geomorphic Hazards and Consequent Risk in Temperate Himachal Himalaya Mapping Scale 1:50,000, In I. Mohan *The Fragile Environment*, New Environment Series, Ashish, New Delhi.

Pirazizy, A.A. 1991 : Demures and Exigency of Deterioration Appraisal on Temperate Soils : Probing Profoundity, In B. Thakur (ed.), *Perspectives in Resource Management in Developing Countries*, Concept's International Series in Geography, Concept, New Delhi.

Pirazizy, A.A. 1991 : Environmental Sustainability and Agricultural Growth, North West Zone India, Sustainability and Agricultural Developpment : A Preliminary Attempt in Modelling, North West Zone India, In Chaudhri, D.P., Singh, R.B., Pirazizy, A.A. and Deb, S. (eds.), *Landuse, Sustainability and Agricultural Development in North West India (1951-88)*, Final Draft, June 1991, Ministry of Agriculture, Government of India.

Pirazizy, A.A. 1991 : Physiography and Soil Resources of North West Zone, India, In Chaudhri, D.P., Singh, R.B., Pirazizy, A.A., and Deb, S. (eds.), *Landuse, Sustainability and Agricultural Development in North West India (1951-88)*, Final Draft, June 1991, Ministry of Agriculture, Government of India.

Pirazizy, A.A. 1991 : Anthropo-Environmental Interactive Systems, Rapid Mobility and Landslides Conundrums of Highlands, In M. Hussain, *Environmental Impact Assessment (EIA)*, Rima New Delhi.

Pirazizy, A.A. 1991 : Environmental Sustainability, Dwindling Ecosystem Energetics and Human Sensibility : Assessment of Change in Temperate

Forest, In I. Mohan, *The Fragile Environment*, New Environment Series, Ashish, New Delhi.

Planning Commission 1982a : *Report of the Task Force for the Study of Ecodevelopment in the Himalayan Region*, Govt. India, pp.152.

Puri, G.S. 1960 : *Indian Forest Ecology*, Vol. I, Oxford Book, New Delhi.

Quiggin, J. 1987 : Land Degradation : Behavoural Causes, In A. Chishotme and Dumsbay, R. (eds.), *Land Degradation Problems and Policies*, Cambridge Univ. Press, Cambridge, pp.203-212.

Raju, K.C.C. and Jalote, P.M. 1980 : Geotectonical Analysis of the Naina Devi Landslide, HP, India, *Geol. Sur. Ind.*, Lucknow, pp.9-12.

Ramsay, W.J.H. 1985 : *Erosion in the Middle Himalaya, Nepal : A Case Study of the Phewa Valley*, Univ. British Columbia, Vancouver.

Randhawa, M.S. 1974 : *Travels in Western Himalayas*, Thimpu, Bhutan.

Rattan, S.S. 1974 : On the Manjir Conglomerate of the Chamba Area, Western Himachal Pradesh, *Himalayan Geology*, 4, pp. 430-446.

Rattan, S.S. 1978 : Chamba Formation, the Late Pre-Cambrian Flysch of the Tethys Himalayan Zone in the Chamba Area, Western Himachal Pradesh, *Bull. Ind. Geol. Assoc.*, 11(2), pp.153-164.

Rattan, S.S. 1979 : On the Kalhel Limestone of the Chamba Area, Western HP, *Bull. Ind. Geol. Assoc.*, 12(1), pp.73-79.

Rattan, S.S. 1984 : Flysch Sedimentation in the Chamba Area, Western Himachal Pradesh, In Srivastava, R.A.K. (ed.), *Sedimentry Geology of the Himalaya*, Today and Tomorrow's Printers and Pub., New Delhi.

Rau, M.A. 1974 : Vegetation and Phytogeography of the Himalaya, In Mani, M.S. (ed.), *Ecology and Biogeogrnphy in India*, Dr. W. Junk, B.V., the Hague, pp.120-152.

Raychaudhri, S.P. 1966 : *Land and Soils*, National Book Thrust, New Delhi.

Robinson, G.W. 1932 : *Soils, their Origin, Constitution and Classification*, London.

Sahu, D. 1990 : *Land Forms, Hydrology and Sedimentation*, Naya Prakash, Calcutta.

Scheidegger, A.E. 1984 : A Review of Recent Work on Mass Movements on Slopes and Slopes or Rockfalls, *Earth Science Review*, 21, p.225-249.

Scheidegger, A.E. 1988 : Hazards from Mass Movements in Mountain Regions, In M.I. Elsabh and T.S. Murty (eds.), *Natural and Man-Made Hazards*, Reidel Pub., Holland, pp.21-41.

Schumacher, E.F. 1975 : *Small is Beautiful*, Harper and Row, New York.

Schuster, R.L. and Krizek, R.L. 1984 : *Landslides : Analysis and Control, Transport Reasearch Board Special Report, 176*, Nat. Acad. Sci., Washington D.C.

Selby, M.J. 1985 : *Earth's Changing Surface*, OUP, Oxford.

Sharma, K.K. 1979 : A Contribution to the Geology of the Sutlej Valley, Kinnaur, Himachal Pradesh, *Colloqu, Int. Du. C.N.R.S.*, 268, pp.369-378.

Sharma, K.K. 1982 : The Sequences of Phased Uplift of the Himalaya, In R.O. White (ed.), *The Evolution of East Asian Environment*, Hong Kong Univ., Hong Kong.

Sharma, K.K. 1983 : The Sequences of Phased Uplift of the Himalaya, In R.O. White (ed.), *The Evolution of the East Asian Environment*, Hong Kong Univ. Press, pp.40-45.

Sharma, V.P. 1975 : On the Age and Lateral Gradation of Seawa Para Gneiss, *Bul. Ind. Geol. Assc.*, Vol. 8, No. 2, pp.113-117.

Shresta, T.B. 1977 : Natural Vegetation of Karnali Zone (North West Nepal) and the People of the Region, In Acto die Colleque International, *Himalaya Ecologie-Ethnologie*, No. 268 edition, DN, CIVRS, Paris, p.155.

Sidle, R.C. et al, 1985 : *Hillslope Stability and Landuse*, Am. Geophys. Union, Washington D.C.

Singh, B. 1952 : *Final Report of the 4th Revised Settlement of Kulu.*

Singh, D.R. and Gupta, P.N. 1982 : Assessment of Siltation in Tehri Reservior, *Proc. Int. Symp. Hydrological Aspects of Mountain Watersheds*, Univ. of Roorkee, Mangalik Prakashan, Saharanpur, pp.60-66.

Singh, J.S. 1985 : *Environmental Regeneration in Himalaya, Concepts and Strategies*, CHEA, Nanital.

Singh, R.L. 1971 : *India A Regional Geography*, Nat. Geog. Soc. Ind. Varanasi.

Singh, T. 1982 : *Studies in Himalayan Ecology and Development Strategies*, The English Book Store, New Delhi.

Singh, U. 1903 : *Conquest of Arki by the Gorkhas*, Lahour Govt. Press.

Skempton, A.W. and Hutchinson, J.N. 1969 : Stability of Natural Slopes and Embarkment Foundations, *State of the Art,* Vol. VII, *Int. Conf. Soil Mech. Fond. Eng.*, Mexico, pp.291-340.

Smith, R.M. 1965 : Determining the Range of Tolerable Erosion, *Soil Science*, 100(5), pp.414-424.

Srikantia, S.V. and Sharma, R.P. 1969 : Soil Formation–A Note on the Stratigraphic Sequence, *Bull. Geol. Soc. Ind.*, 6(3), pp.93-97.

Srivastava, R.N. and Sharma, O.P. 1975 : Report on Geological Mapping in Parts of Kangra and Kulu Districts, HP, *Himalayan Geology*, (4), pp.369-413.

Strachey, H., 1859 : *Physical Geography of Western Tibet*, London.

Sub-Division of the Kangra District 1945-52, Govt. Press Chandigarh.

Tejwani, K.G. 1987 : Water Management in the Indian Himalaya, In T.N. Khoshoo (ed), *Perspective in Environmental Management*, Oxford and IBH, New Delhi, pp.203-227.

Thompson, T. 1852 : *Western Himalayas and Tibet : The Mountains of North India During 1847-48*, London.

Tolba, M.K. 1986 : *Development without Destruction : Volving Environmental Perceptions*, Tycooly, London.

Trevor, C.G. 1920 : *Revised Working Plan for the Kullu Forests 1919-20 to 1943-44*, Lahour, Government Press.

Trowbridge, C.C. 1913 : On Fundamental Methods of Orientation and "Imaginary Maps", *Science*, 88, pp.888-896.

Trudgil, S.T. 1987 : Soil Profile Processes, In K.J. Gregory and D.E. Walling (eds.), *Human Activity and Environmental Processes*, Wiley, London, pp.307-327.

Turk, J. and Turk, A. 1988 : *Environmental Science*, Saunder Co., New York.

UNESCO 1982 : Sedimentation Problems in River Basins, *UNESCO*, Paris.

Valdiya, K.S. 1973 : Tectonic Framework of India : A Review and Interpretation of Recent Structural and Tectonic Studies, *Geophys, Res. Bull.*, Vol. 11, No. 2, pp.79-114.

Valdiya, K.S. 1987 : *Environmental Geology : India Context*, Tata McGraw-Hill, New Delhi.

Varnes, D.J. 1958 : Landslide Types and Processes, E.B. Eckel (ed), Landslides and Engineering Practice, *Highway Research Board Washington Spec. Rep. 29; NAS-NRC Publ. 544*, pp.20-47.

Varnes, D.J. 1978 : Slope Movements and Types and Processes, in *Landslides : Analysis and Control, Spec. Rep. 176*, Transportation Res. Board Nat. Acad. Sci., Washington, pp.11-33.

Varnes, D.J. 1984 : *Landslide Hazard Zonation : A Review of Principles and Practice*, UNESCO, Paris.

Verent, A. 1958 : Arid Zone Reasearch, UNESCO, X.75, In Griffiths, J.F. (ed), *Applied Climatology an Introduction*, Oxford Univ. Press, London, pp.55-58.

Veder, C. 1979 : *Landslides and their Stabilization*, Springer Verlong, New York.

Vink, A.P.A. 1975 : *Landslides in Advancing Agriculture*, Springer Verlag, Berlin.

Voight, B. 1978 : *Rockslides and Avalanches I : Natural Phenomena*, Elsevier, Amsterdam.

Vuichard, D. and Zimmermann, M. 1986 : The Langmoche Flash Flood, Khumbu Himal, Nepal, *Mountain Research and Development*, 6(1), pp.90-94.

Wadia, D.N. 1931 : The Syntaxis of North-Western Himalaya Tectonics and Orogeny, *Rec. Geol. Surv. Ind.*, (65), pp.189-220.

Wadia, D.N. 1937 : The Tectonics of North India, *Proc. Int. Geol. Congr. 17th Sess*. Moscow, Vol. II, pp.4343-4360.

Wadia, D.N. 1949 : *Geology of India*, London.

Walling, D.E. 1987 : *Human Activities and Environmental Processes*, Wiley, New York.

Walling, D.E. 1987 : Hydrological Process, In K.J. Gregory and Walling, D.E. (eds.), *Human Ativity and Environmental Processes* Wiley, Chichester, pp.53-86.

Watson, J. 1983 : *Geology and Man an Introduction to Applied Earth Sciences*, George Allen and Unwin, London.

Webb III, T. et al 1985 : 20000 Years of Global Climatic Change, In T.F. Malone and J.G. Roederer (eds.), *Global Change*, Cambridge Univ. Press, New York, pp.184-185.

West, R.G. 1964 : Inter-Relations of Ecology and Quarternary Palaeobotary, In T.F. Malone, and J.G. Roederer (eds.), *Global Change*, Cambridge Univ. Press, New York, pp.105-108.

White, G. 1945 : *Human Adjustments to Floods*, Univ. of Chicago, Chicago.

White, G.F. 1974a : Natural Hazard Research : Concepts, Methods and Policy Implications, In White, G.P. (ed), *Natural Hazard : Local National*, Global, New York, pp.3-16.

Whyte, A.V.T. 1986 : From Hazard Perception to Human Ecology, In R.W. Kates and Burton, I. (eds.), *Geography, Resources and Environment*, Vol. III, Univ. of Chicago, pp.240-271.

Williams, P.J. 1982 : *The Surface of the Earth : An Introduction to Geotechnical Science*, Longman, London.

Winter, E.J. 1974 : *Water, Soil and the Plant*, MacMillan, London.

Wischmeier, W.H. 1959 : Rainfall Erosion Index for a Universal Soil-Loss Equation, *Soil Sci. Soc. Amr. Proc.*, (23) pp.246-249.

Wischmeier, W.H., Smith, D.D. and Uhland, R.E. 1958 : Evaluation of Factors in the Soil-Loss Equation, *Agricultural Engineering*, 39(8), pp.358-359.

Wolman, M.G. and Schick, A.P. 1967 : Effects of Construction on Fluvial Sediment, Urban and Suburban Areas of Maryland, *Water Resources Res.*, No. 3, pp.451-464.

Wolpert, J. 1964 : The Decission Process in a Spatial Context, *Annl. Assoc. Amr. Geog.*, 54, pp.536-558.

Wright, J.R., 1947 : Terre Incognitae : The Place of Imagination in Geography, *Annl. Assoc. Amr. Geog.*, 37, pp.1-15.

Young, A. 1960 : Soil Movement by Denudation Processes on Slopes, *Nature*, 188, pp.120-122.

Zachar, D. 1982 : *Soil Erosion*, Elsevier, Amesterdam.

Zaruba, Q. and Mencl, V. 1969 : *Landslides and Their Control*, Elsevier, Amsterdam.

Zdrazil, K. 1970 : Economic Evaluation of Erosion Control (Method of Calculation), In Stehlik, O. (ed.) *Geographical Regionalization of Soil Erosion in C.S.R. Methods of Elaboration*, CSAV Pub. Prague.

Zeitler, P.K. et al, 1982 : Fission-Track Evidence for Quaternary Uplift of the Nanga Parbat Region Pakistan, *Nature*, 298, pp.255-257.

Zimmermann, M. et al 1986 : Mountain Hazard Mapping in the Khumbu Himal Nepal, with Prototype map, scale 1:5000, *Mountain Research and Development*, 6(1), pp.29-40.

INDEX